W0111653

Water in Mineral Processing

Water in Mineral Processing

Editor

Omvir Singh

Water in Mineral Processing

Edited by **Omvir Singh**

Printed in 2017

ISBN: 978-1-68117-498-3

Library of Congress Control Number: 2015936602

© 2016 by
SCITUS Academics LLC,
616, Corporate Way, Suite 2, 4766,
Valley Cottage, NY 10989

www.scitusacademics.com

This book contains information obtained from highly regarded resources. Copyright for individual articles remains with the authors as indicated. All chapters are distributed under the terms of the Creative Commons Attribution License, which permits unrestricted use, distribution, and reproduction in any medium, provided the original author and source are credited.

Notice

Reasonable efforts have been made to publish reliable data and views articulated in the chapters are those of the individual contributors, and not necessarily those of the editors or publishers. Editors or publishers are not responsible for the accuracy of the information in the published chapters or consequences of their use. The publisher believes no responsibility for any damage or grievance to the persons or property arising out of the use of any materials, instructions, methods or thoughts in the book. The editors and the publisher have attempted to trace the copyright holders of all material reproduced in this publication and apologize to copyright holders if permission has not been obtained. If any copyright holder has not been acknowledged, please write to us so we may rectify.

Contents

Preface

Water in Mineral Processing is an authoritative, first-of-its-kind resource that can help mining practitioners apply innovative water-use and purification technologies in the demanding years ahead. Water in Mineral Processing provides a comprehensive, state-of-the-art examination of this vital issue. A compilation of papers presented at the First International Symposium on Water in Mineral Processing, this book shares the insights of dozens of respected experts from industry and academia. A significant portion of the content is devoted to saline solutions and processing with sea water.

Editor

Assessment of Natural Uranium in the Ground Water around Jaduguda Uranium Mining Complex, India

N. K. Sethy[1], R. M. Tripathi[2], V. N. Jha[1],
S. K. Sahoo[2], A. K. Shukla[2], and V. D. Puranik[2]

[1]Environmental Assessment Division, Bhabha Atomic Research Centre, Health Physics Unit, Jaduguda Mines, Jharkhand, India

[2]Bhabha Atomic Research Centre, Mumbai, India

ABSTRACT

Ground water ecosystem surrounding the uranium processing facility at Jaduguda, India has been studied for natural uranium distribution. Annual intake of uranium through drinking water for members of public residing around the uranium complex is found to be in the range of 41.8 $Bq \cdot y^{-1}$ - 44.4 $Bq \cdot y^{-1}$. The intake and ingestion dose is appreciably low (<2 $\mu Sv \cdot y^{-1}$) which is far below the WHO recommended level of 100

$\mu Sv \cdot y^{-1}$. The excess life time radiological risk due to uranium natural in drinking water is insignificant and found to be of the order of 10^{-6}. Even the highest concentra-tion of uranium was found to be 28 $\mu g \cdot l^{-1}$ is away (at 1.5 to 5 km distance) from mining industry and well below the acceptable limit. The ground water in the area around the uranium facility is not affected by the mining activity. The ground water in three zones is safe and reflects the natural distribution of uranium.

INTRODUCTION

General Description

Rapid industrialization and subsequent waste disposal has been a concern for ground water contamination. Ground water is the major source of drinking water in many parts India. Industrial activities, metal mining and waste depositary may contribute to the nearby ground water sources by radionuclide migration. Mining and processing of uranium in the east-Singbhum region of Jharkhand has been started in early sixties. Uranium ore is mined from a cluster of mines (Jaduguda, Bhatin and Narwapahar, Turamundih, Bagjata, Mauldih) spreading in the region and processed at centralized ore-processing plant at Jaduguda using hydrometallurgy technique. As the ore is of very low grade (<0.05% U_3O_8), a huge quantity of process waste (tailings) is generated and dis-posed safely in tailings ponds. Tailings pond is an engineered having valley with natural hills on three sides and earthen bund forming the fourth side. Engineering features of the earthen bund ensures the decantation of dis-solved radionuclides, which are treated further for removal of the toxins (U, ^{226}Ra and heavy metals) prior to their discharge into the aquatic ecosystem. The change in physicochemical characteristics of tailings over the period may take place leading to dissolution of some of the contaminants. The migration of these contaminants into the adjoining ground water sources can be anticipated. Evaluation of ingestion dose and subsequent risk due to intake of water to population residing around the tailings pond is the subject matter of this study. The mining complex comprises of uranium mines, ore processing plant and tailings ponds. The study area is situated at Jaduguda (22°30′N and long. 85°40′E) in the East Singbhum district of Jharkhand, India. The area is well known for its wide mineral deposits and receives >1000 mm of rain fall annually. The maximum temperature in summer is >45°C and minimum is <7°C during winter.

Health Hazard of Uranium

Toxicity of uranium has been established by animal studies and human data from uranium miners and workers with accidental exposures indicate that uranium affects the proximal tubules of the kidney; at very high acute doses, tubular degeneration and necrosis (that is, death of tissue) may occur a few days after the intake of uranium [1]. Kidney is generally considered to be the critical organ for uranium through water or food. The uranyl ion forms bicarbonate, citrate and $UO_3(CO_2)_3$ complexes in blood plasma [2]. The UO_2^{++} ion binds with the red blood cells. During the purification of blood in kidney, it is filtered from the blood and then recombines with the cell surface ligands. Studies on uranium toxicity studies in human have been described elsewhere [3,4]. Though, intake of uranium by members of the public can occur through various routes. However the principal route of ingestion of uranium is through drinking water [5] and to a lesser extent through the foodstuff. Intake of uranium through drinking water by population residing around the uranium mining area has been considered in the present study. United States Environmental Protection Agency (EPA) has classified uranium as a group- A human carcinogen. It has prescribed maximum contaminant level goal (MCLG) for uranium as 0 (zero) in 1991(zero tolerance). In drinking water, EPA suggests maximum contaminant level as (MCL) of 30 $\mu g \cdot l^{-1}$ [6]. In Canada, the proposed interim maximum acceptable level is (IMAC) of 20 $\mu g \cdot l^{-1}$, whereas World Health Or-ganization (WHO) strictly recommended a reference level as 2 $\mu g \cdot l^{-1}$ [7].

MATERIALS AND METHODS

Sampling and Analysis

Grab samples (5 lit) were collected from open wells and tube wells situated at various distances in the public do-main. The area around the uranium mining Industry is divided into three zones *i.e.* <1.5 km, 1.5 to 5 km and >5 km. Ground water samples were collected from these three zones. Sample locations were selected covering three measure season of the area (October to March, April to June and August to September) on the basis of public utility and down stream direction from uranium industry. More number of samples was collected from downstream side of the Uranium Industry. Samples were brought to the laboratory filtered and preserved in acidic medium. About 100 ml of water sample is evaporated to dryness and 20 ml of 0.25 N electronic grades pure H_2SO_4 is added and reflux for 30 minutes in a hot plate. It is then cooled and transferred to a separating funnel/tube. Then 20 ml of alamine-benzene (2% alamine in 98% benzene) solution is added and the mixture is shaken for few minutes with occasionally opening the mouth of the separating tube to vent off the gases formed inside. The aqueous phase is drained out and 0.1 ml from organic phase is taken for planchatting in a platinum disc. The basic principle of estimation of natural uranium in envi-ronmental sample is to quantitatively transfer the trace amount of uranium present in the sample aliquot to a small platinum disc and measure the intensity of flores-cence of uranium compound. A small volume (0.1 ml) of organic media containing the uranium is transferred to platinum disc, fused with 250 mg of NaF-Na_2CO_3 (15:85) fusion mixture at 800°C for 3 minutes. Cooled and fluo-rescence intensity was measured in ECIL, make Fluori- meter (Model No: FL6224A) [8].

UV radiation of excitation wavelength 3650Å is irra-diated on the platinum disc containing fused sample and emitted florescence of 5546Å wavelength is unique to uranium [9]. Intensity of fluorescence is proportional to the amount of uranium present in the sample. Standard (1 ug/ml) and blank were processed simultaneously and uranium was estimated by using the formula

$$U(\mu g) = \frac{\text{Sample reading} - \text{Blank reading}}{\text{Standard reading} - \text{Blank reading}}$$

The uranium content of the original sample was ob-tained from the above equation by further applying the sampling parameters. The advantage of this method is aqueous to organic ratio is not critical and may be varied over a wide range. There is no need of any salting out reagent and organic phase can directly planchatted. The disadvantage of this method is that the fusion mixture is hygroscopic and measurement of fluorescence intensity should be measured without giving much delay. The de-tection limit of this method is 0.1 μg.

Quality Control

Quality assurance of the analytical procedure followed was determined by using certified reference materials (SRMs) produced by Canada Center for Mineral and Energy Technology (CANMET) and supplied by Bureau of analysed samples Ltd., UK. Environmental reference sample such as lake sediments (LKSD-1&3), stream sediment (STSD-1) and geochemical soil (TILL-1&2) were analysed for natural uranium. The result of the analysis is presented in Table 3.

RESULT AND DISCUSSIONS

Distribution of Uranium in water and Intake

The histogram of uranium concentration in ground water around the uranium mining industry over the study pe-riod is presented in Figure 1. A year wise geometric mean concentration of uranium in the ground water sources in different distance zones is presented in Table 1. However, the maximum concentration of U(nat) within 1.6 km distance was observed at 11 $\mu g \cdot l^{-1}$ with GM of 1.13 $\mu g \cdot l^{-1}$ and GSD of 2.64 during the entire study period. The large variation in the uranium concen-tration is due to uneven distribution of uranium in the lithosphere. Further, the histogram plot (Figure 2) of the data set reveals that in majority of ground water source the concentration was less than 0.5 $\mu g \cdot l^{-1}$. The distribution of U(nat) during the study period in the distance zone 1.6 km - 5 km was varied with GM concentration of 1.2 mg.m-3 and GSD of 3.4. The maximum concentration in this zone was 28 $\mu g \cdot l^{-1}$. In the case of natural unmined ore deposits uranium can enter the ground water by way of leaching of uranium bearing rock strata by the ground water aquifers. The physicochemical environment around the source has great influence on distribution of uranium natural in ground water. In this context solubility of uranium in the medium is probably playing a vital role. Only the hexavalent uranium compounds are soluble which is favored by aerobic condition of the environment.

The less soluble tetravalent fraction can get dissolved and variation in the levels can be expected even within the same geological formations. Apart from this pH, competing ions, complex formation with uranyl ions, seasonal variations are also leading to variable distribution of uranium in the ground water. During the study period, in the distance zone of >5 km, distribution of U(nat) was varied with GM of 1.13 $\mu g \cdot l^{-1}$ and GSD of 2.28. One way ANOVA (Table 2) reveals that there is insignificant variation in uranium natural concentration at different distances from the tailings pond with Chi-square > p 0.052. The maximum concentration within a distance of 1.6 km was appreciably low as compared to the recom-mended national regulatory standard of 60 $\mu g \cdot l^{-1}$ based on the F class radiological consideration. Since uranium appearing in drinking water is soluble uranium the most restrictive radiological class has been considered for recommending the limits. In other zone also the recom-mended limit of 60 $\mu g \cdot l^{-1}$ has not been exceeded so far. The significantly lower concentration within a distance of 1.6 km from uranium industry is attributed to the local geological features of the area. This also confirms that so far there has been insignificant migration of U(nat) from the operation of uranium industry to the adjoining ground

water sources. The present study is compared with simi-lar studies carried out in other countries (Table 5) as well as in India. The concentration range in ground water in the present study, as evident in Table 5 is comparable to study carried out in Turkey [10] but lower than similar studies in USA [11,12].

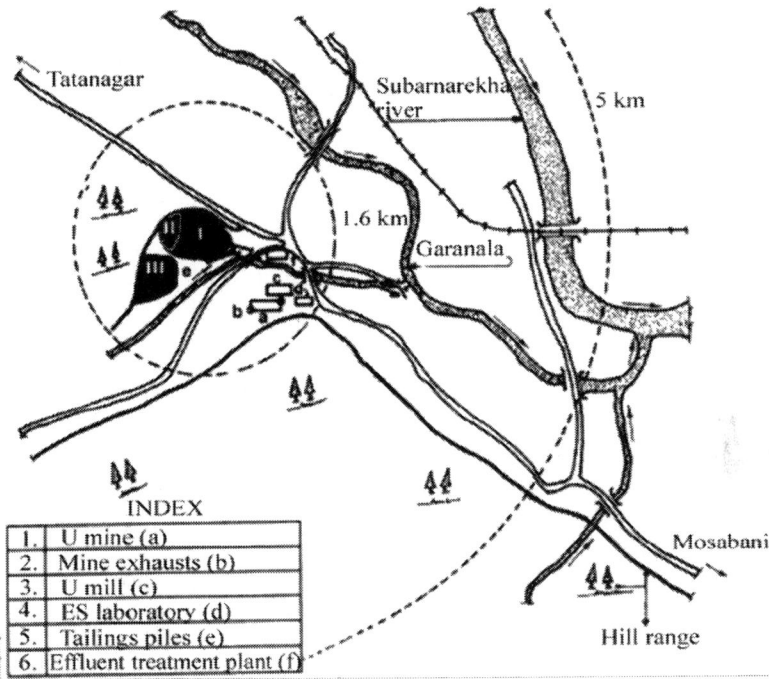

Figure 1: The Environmental map of Uranium Mining Complex, Jadugoda, India.

Table 1: Concentration of U(nat) in ground water samples in various distance zones

Distances from T.P (km)	No of samples (N)	Year	U(nat) ($\mu g \cdot l^{-1}$)	Geo Mean (GSD)
		2003	Range	
<1.6	8		0.5 - 11	1.62(3.1)
1.6 - 5.0	17		0.5 - 20	1.21(3.2)
>5.0	20		0.5 - 4.7	1.0 (2.5)
		2004		

<1.6	15		0.5 - 7.6	1.39(3.0)
1.6 - 5.0	15		0.5 - 28	2.8(3.7)
>5.0	10		0.5 - 4.3	2.79(3.5)
		2005		
<1.6	16		0.5 - 7.6	1.31(2.9)
1.6 - 5.0	16		0.5 - 19	2.39(3.2)
>5.0	11		0.5 - 4.3	1.49(2.2)
		2006		
< 1.6	5		0.5 - 2.3	0.86(2.1)
1.6 - 5.0	4		0.5 - 1.2	0.76(1.45)
>5.0	17		0.5 - 4.7	1.37(3.9)
		2007		
<1.6	15		0.5 - 1.5	0.71(1.5)
1.6 - 5.0	9		0.5 - 3.1	1.0 (1.9)
>5.0	10		0.5 - 3.5	0.78(1.9)

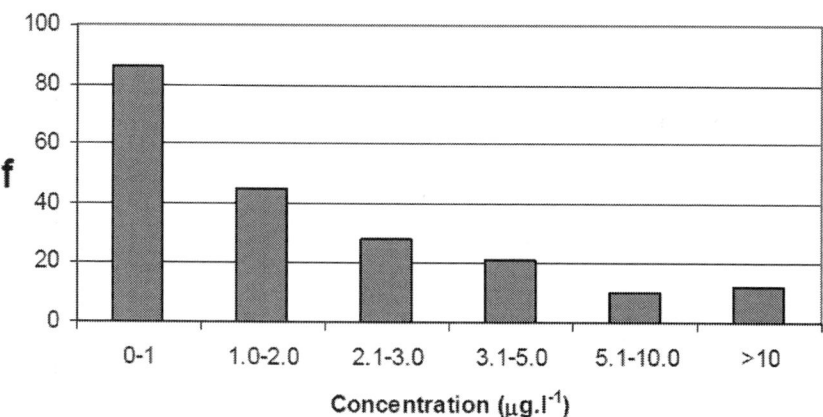

Figure 2: Histogram of U in Ground water.

Radiation Dose and Risk

A five year study of ground water as discussed earlier showed natural uranium was in low concentration range. Though mass concentration of natural uranium was measured in ground water samples it can be expressed in activity concentration using conversion factor of 25 mBq·µg^{-1}. Uranium level (Bq·m^{-3} or µg·l^{-1}) = Measured mass concentration (µg·l^{-1}) × Conversion factor (25 mBq·µg^{-1}) Considering the daily intake of water for In-dian reference man 4.05 l·d^{-1} [13,14], dose conversion factor of 0.045 µSv·Bq^{-1} [15] the annual ingestion dose to the adult individual residing around the tailings pond due to natural uranium was estimated. Since the data distribution in the three zones can be approximated by log normal the intake and ingestion dose should be based on geometric mean. Accordingly the annual intake from drinking water in the three zones discussed can be esti-mated to be 41.8, 44.4 and 41.8 Bq·Y^{-1} with an ingestion dose of 1.88, 2.0 and 1.88 µSv·Y^{-1} respectively. This is far lower than the [16] recommended guideline of 100 µSv·Y^{-1} for ingestion from the intake of a single ra-dionuclide. Health effects due to exposure of uranium can be classified as radiological risk as radioactive ele-ment and chemical risk as a heavy metal. Radiological risk was evaluated using the risk coefficient 4.40 × 10-11per pCi as per the US EPA [17] standard method. The radiological risk was converted to excess lifetime risk by multiplying with activity concentration of ura-nium level in ground water. While estimating the risk average body weight (52kg) and water intake by Indian reference man (4.05 liter per day) were considered. The average life expectancy in India is 60 years was consid-ered as total exposure period. Finally the radiological risk coefficient was estimated as 1.06 × 10^{-4} in the Indian scenario.

The illustrative calculation is:

Risk Factor (per Bq·l^{-1}) = Risk coefficient (4.40 E - 11) × water ingestion rate (4.05 l·day^{-1}) total exposure dura-tion (21,900 day) × conversion factor (27 pCi·Bq^{-1}) = 1.06 × 10^{-4} Bq·l^{-1}.

Excess life time cancer Risk = uranium level (Bq.l^{-1}) × Risk factor (1.06 × 10^{-4} Bq·l^{-1}).

The radiological risk estimated is presented in Table 4. The risk was found to be in the order of 10^{-6} and much below the acceptable radiological risk of 10^{-3} [7]. Hence, the radiological risk due to natural uranium in ground water might be acceptable and the uranium mining in-dustry has insignificant impact in the ground water ura-nium concentration of the study area.

Table 2: Kruskal-Wallis One-Way ANOVA for U(nat) variation around tailings pond

Source	SS	df	MS	Chi Square	Prot) > Chi-Square
Distance	10898.2	2	54449.1	5.89	0.052
Error	270313.3	150	1802.09		
Total	281211.5	152			

Table 3: Concentration of Uranium in certified reference materials

Reference Material	Unit	Certified Concentration	Observed Concentration
LKSD1	$\mu g\, g^{-1}$	9.7	8.8 ± 0.4
LKSD 3	$\mu g\, g^{-1}$	4.6	4.4 ± 1.2
TILL 3	$\mu g\, g^{-1}$	2.2	1.9 ± 0.6
STSD-1	$\mu g\, g^{-1}$	2.1	1.8 ± 0.5
STSD-1	$\mu g\, g^{-1}$	8.0	7.8 ± 1.3

Table 4: Excess Radiological Risk for ground water Uranium at various distance regions (2003-07)

Distance	U(mg·m⁻³) Geo mean	Excess Radiological Risk
<1.5 km	2.29	6.06×10^{-6}

1.5 km - 5 km	1.62	4.31×10^{-6}
>5 km	1.17	3.11×10^{-6}

Table 5: Comparison of uranium concentration in drinking water in different countries

Country	Range of U(μg•r^i)	References
Turkey	0.24 - 17.65	Kurnru *at al. [10]*
South Greenland	0.5 - 1.0	Brown *et al* [18]
USA	0.01 - 652	Cothern *et al* [11]
Kuwait	0.02 - 2.48	Bou-Rabee *et al* [12]
Jordan	0.04 - 1400	Smith *et al* [19]
Central Austrlia	>20	Hostetler *et al* [20]
Cochin, India	0.34 - 2.54	Prabhu R.S. *et al* [21]
Jaduguda, India	0.50 - 28	Present Study

CONCLUSIONS

The distribution of U(nat) in ground water reflects the natural background of the area. The variation in concen-tration at different distances may be attributed to the geological features of the area and the physicochemical environment around the source. A slight reduced ura-nium concentration in the <1.5 km or closest to uranium industry may be attributed to the soil/rock type around the ground water sources. The highest concentration of uranium was found to be 28 μg·l^{-1} is away (1.5 to 5 km distance zone) from mining industry. The intake and in-gestion dose is appreciably low (2 µSv·Y^{-1}) which is far below the WHO recommended level of 100 µSv·Y^{-1}. The risk due to radiological is acceptable and very low. It can be concluded that the ground water around the uraniumindustry at Jaduguda is not affected by the uranium min-ing activity. The drinking water studied in three different distance zones of uranium mining facility found to be safe.

REFERENCES

1.   "Ask the Expert," Health Physics Society, April 2002. http://www. hps.org/publicinformation/ate/q1906.html

2.   P. W. Durbin, "Metabolic Model for Uranium," In: R. H. Moore, Ed, *Biokinetics and Analysis in Man* 1984, United States Uranium Registry, National Technical Information Service, Springfield, 1984.

3.   J. B. Hursh and N. L. Spoor, "Data on Man," In: H. C. Hodge, J. N. Stannard and J. B. Hursh, Eds., *Uranium, Plutonium, Transplutonium Elements*, Springer-Verlag, Berlin, 1973, pp. 197-239.

4.   A J. Lussenhop, J. C. Gallimore, W. H. Sweet, E. G. Struxness and J. Robinson, "The Toxicity in Man of Hexavalent Uranium Following Intravenous Administra-tion," *American Journal of Roentgenology,* Vol. 79, No. 1, 1958, pp. 83-100.

5.   S. C. Morris and A. F. Meinhold, "Probabilistic Risk Assessment of Nephrotoxic Effect of Uranium in Drink-ingWater," *Health Physics,* Vol. 69, No. 6, 1995, pp. 897-908. doi:10.1097/00004032-199512000-00003

6.   U.S. EPA, Draft Guidelines for Carcinogen Risk Assess-ment (Review Draft, July 1999), U. S. Environmental Protection Agency, Risk Assessment Forum, Washington, D.C., 1999.

7.   D. C. Shin, Y. S. Kim, J. Y. Moon, H. S. Park, J. Y. Kim and S. K. Park, "International Trends in Risk Manage-ment of Groundwater Radionuclides," *Journal of Envi-ronmental Toxicology,* Vol. 17, No. 4, 2002, pp. 273-284.

8.   K. P. Eappan and P. M. Markose, "Amine Extraction for Uranium Estimation," *Bulletin of Radiation Protection,* Vol. 9, 1986, pp. 83-86

9.   M. Kolthoff and P. J. Eiving, "Treatise on Analytical Chemistry, Part II. Vol. 9," Wiley, New York, 1962, pp. 102-111.

10. M. N. Kumru, "Distribution of Radionuclides in Sedi-ments and Soils along the Buyuk Mendres River," *Pro-ceeding of Pakistan Academy of Sciences*, Vol.32, 1995, pp. 51-56

11. C. R. Cotheren and W. L. Lappenbusch, "Occurance of Uranium in Drinking Water in the US," *Health Physics*, Vol. 45, No. 1, 1983, pp. 89-99. doi:10.1097/00004032-198307000-00009

12. F. Bou-Rabee, "Estimating the Concentration of Uranium in Some Environmental Samples in Kuwait after the 1991 Gulf War," *Applied Radiation and Isotopes*, Vol. 46. No. 4, 1995, pp. 217-220. doi:10.1016/0969-8043(94)00122-G

13. B. M. Raghunath and S. D. Soman, "Water Intake Data for Indian Reference Man," *Indian Journal of Environ-mental Health*, Vol. 2, 1969, pp. 1-7.

14. H. S. Dang, D. D. Jaiswal, M. Parameswaran and S. Krishnamony, "Physical, Anatomical, Physiological and Metabolic Data for Reference Indian Man—A Proposal," Bhabha Atomic Research Centre, Mumbai, 1994.

15. "Dose Coefficients for Intakes of Radionuclides by Workers," International Commission on Radiological Protection Series, Annals of the ICRP, Elsevier Health Sciences, Amsterdam, 1995.

16. WHO, Guidelines for Drinking-water Quality, Volume 1, Recommendations, 2004.

17. SEPA, National Primary Drinking Water Regulation Ra-dionuclide, Final Rule, 2000.

18. A Brown, A. Steenfelt and H. Kunzennorf, "Uranium Districts Defined by Reconnaissance Geo Chemistry in South Greenland," *Journal of Geochemical Exploration*, Vol. 19, No. 1-3, 1983, pp. 127-145. doi:10.1016/0375-6742(83)90013-4

19. B Smith, A. E. Powel, A. Milodowski, *et al.*, "Identifica-tion, Investigation and Remediation of Ground Water Containing Elevated Level of Uranium-Series Radionu-clides: A Case Study from the Easteron Mediterranean," *Proceddings of the 3rd International Conference on the Geology of the Easteren Mediterranean*, Nicosia, Cyprus, 2000.

20. S. Hostetler, J. Wischisen and G. Jacbson, "Ground Water Quality in the Papunya-Kingtore Region Northen Terri-tory," Austrilian Geological Survey Organization, Can-berra, 1998.

21. R. S. Prabhu, R. Sathyapriya, S. K. Sahoo and S. Maha-patra, "Ingestion Dose Due to Natural Uranium to the Public through Drinking Water Pathways in Two Districts of Keral," *Proceedings of 16th National Symposium on Environment*, Hisar, 2008, pp. 551-555.

Chapter 2

Seasonal Variation of the Physicochemical Properties of Water Samples in Mahanadi Estuary, East Coast of India

Pravat Ranjan Dixit[1], Biswabandita Kar[1],
Partha Chattopadhyay[2], and Chitta Ranjan Panda[2]

[1]KIIT University, Bhubaneswar, India
[2]CSIR-Institute of Minerals and Materials Technology, Bhubaneswar, India

ABSTRACT

The two major sources which are contributing to marine pollution are natural processes as well as anthropogenic activities. The natural process includes precipitation, erosion, weathering of crystal material whereas anthropogenic activities are urbanization, industrialization, mining and agricultural activities, etc. Mahanadi is the biggest river source of Odisha which joins the Bay of Bengal at Paradip. Paradip Township is an urbanized well-developed industrial township where various anthropogenic activities are contributing pollution to the water sources. In the present study, an attempt has been made to estimate and monitor the seasonal and spatial variation of physiochemical properties of the Mahanadi estuary, the East Coast coastal belt of India. The result revealed that there is a remarkable variation in the physiochemical parameters such as pH, salinity, TSS, DO, BOD, NO_2N, NO_3N, NH_4N, TN, TP, SIO_4 and Chl-a which are attributed to the runoff water getting discharged to the coastal water sources. It is being recommended to treat waste water and materials before discharging them to the marine water medium.

INTRODUCTION

The marine environment, especially costal and estuary, forms an essential component of the global life. In highly developed countries, all these human activities can affect the characteristics of the water in their estuaries. The environmental impact of municipal wastewater and industrial effluents, discharge on receiving water are numerous and inputs of contaminants can affect the aquatic biota as well as the health of the marine environment [1]. The water quality depends on both natural processes, such as precipitation erosion, weathering of crystal materials and anthropogenic processes like urbanization, industrialization, mining and agricultural activities [2]. These two parameters play a vital role in nutrient cycling, eutrophication, biota abundance and overall food web dynamics in estuarine and near shore ecosystems, whereas surface runoff is a seasonal phenomenon largely affected by climate in the basin. Apart from this, fishing activities near the estuary also influence the water quality. In this context, researchers have analyzed the water quality of Mahanadi River and estuarine system [1,3-7]. The aim of present research is to study the seasonal and special nutrient variation and change in physicochemical parameter due to the impact of pollution sources and industrial effluents upon Mahanadi estuarine environment.

OUTLINES STUDY OF SAMPLING AREA

Mahanadi is one of the major rivers of India and is the largest river of Orissa state having the annual discharge of 66,640 Mm^3. The average annual rainfall is 1572 mm, of which 70% is precipitated during the southeast monsoon between mid June to mid October. The growth of industrialization and urbanization in the upstream and estuarine region of Mahanadi, is putting unique pressure on estuarine and coastal resources [2,5,7]. The Mahanadi River begins in the Baster hills of Madhya Pradesh flows over different geological formations of Eastern Ghats and adjacent areas and joins the Bay of Bengal after divided into different branches in the deltaic area. The main branches of River Mahanadi meet Bay of Bengal at Paradip (Figure 1). The River basin (80°30›E - 86°50›E and 19°20›N - 23°35›N) extends over an area approximately 141,600 km^2, has a total length of 851 km and an annual runoff of 50×10^9 m^3 with a peak dis- charge of 44,740 m3s−1 [8-11]. Mahanadi estuary, the major estuary of coastal Orissa has its drainage basin, which is one of the biggest drainage basins along the east Coast. The tidal estuarine part of the river covers a length of 10 km and has a basin area of 9 km2 [12]. Based on physical characteristics, the estuary has been characterized as a partially mixed coastal plain estuary [13]. Tidal cycle is semidiurnal. It is principally a wave-dominated coast during the southwest monsoon season, while during the non-monsoon period it is mixed wave and tide dominated. Mahanadi estuary receives effluents from two fertilizer plants, other small industries and domestic sewages from Paradip Township. Apart from these, fishing harbour activities at the estuary also affect the water quality. Atharbanki Creek is heavily loaded with the most of the municipal sewage of Paradip Township and effluents from industries. The anthropogenic/port activities influenced the estuarine environment of Mahanadi.

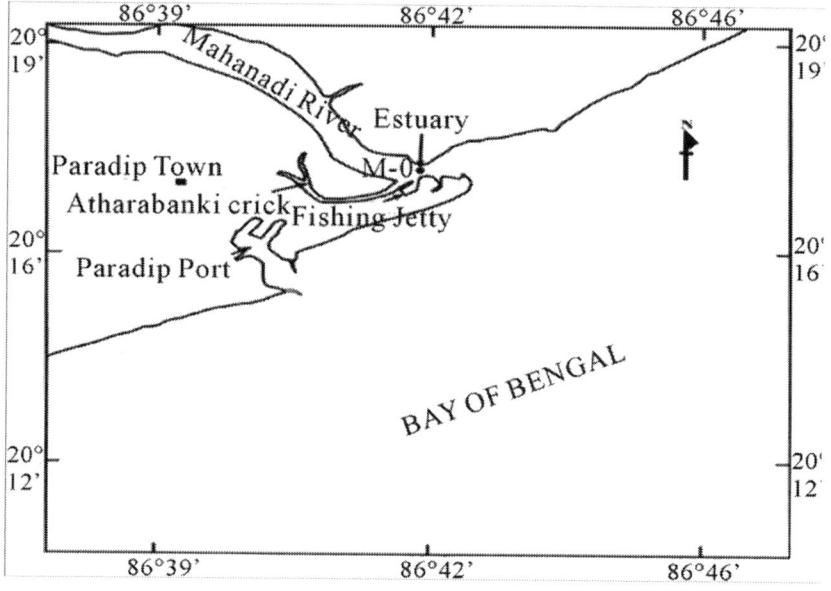

Figure 1: Map showing locations of station in Mahanadi estuary.

MATERIALS AND METHODS

Water samples were collected from the Mahanadi river mouth at a station (Lat. 20°17′16″ and Long. 86°42′28″) during 2008-09 to 2009-10 (Figure 1). Diurnal tidal surface sampling was carried out at an interval of 3 hrs in three different seasons such as, pre-monsoon (May) postmonsoon (October) and summer (March). Surface water temperature, pH were measured in situ by using WTW Kit. Niskin's water sampler will collect water samples. Salinity, TSS, dissolved oxygen (DO) was measured by the standard methods [14]. DO and BOD were done by Winkler's method and BOD was calculated from different in DO concentration after 5 days of incubation at 20°C. Cellulose nitrate membrane filters of pore size 0.45 µm were used for determination of TSS. Nutrients such as NO_2-N, NO_3-N, TN, PO_4-P, TP, and SiO_4-Si were determined by UV-visible spectrophotometer (Perkin Elmer, Lambda 35) as described in methods of seawater analysis [15]. Water samples for Chl-a determination were filtered through Whatmann GF/C glass fiber filters and pigment extraction was performed using

90% acetone. Pigment concentration was measured by means of a UV-visible spectrophotometer for chlorophyll-a analysis [16]. Statistical analysis was carried out by using SPSS VER. 10.0 statistical packages.

RESULT AND DISCUSSION

The seasonal variations of environmental variables are presented in Table 1. The seasonal water temperature varied from 27.18 to 30.85˚C. During higher temperature in summer followed by post monsoon and pre monsoon, the pH varied between 7.99 to 8.43 in pre monsoon, post monsoon and summer. The temperature and pH were recorded lower during the post monsoon seasons. Salinity values fluctuated between 0.01 to 29.36 PSU in premonsoon, post monsoon and summer. The salinity showed a regular trend of variation with tides. The lowest and highest values are always associated with flood and ebb tides. The seasonal variation salinity (0.01 PSU) lowers in Post monsoon, due to the river run off in this period and high dilution in the estuary. The seasonal variation TSS varied from 3.53 mg/l to 56.88 mg/l. The low tide higher value (56.88 mg/l) were recorded during post monsoon, which indicates the river run off, industrial effluents and municipal sewage contaminated in the estuary. The seasonal variation of DO values were between 4.74 mg/l and 8.48 mg/l. The DO values are lowest (4.74 mg/l) in pre monsoon. The lower concentration of DO values in the estuary indicates that a high or-

Table 1: Seasonal variation of standard deviation (Sd.dev.) values of environmental variables in the Mahanadi estuary of year 2008-2010

	PREMONSOON	POSTMONSOON	SUMMER
	Mean ± Sd.dev	Mean ± Sd.dev	Mean ± Sd.dev
pH	8.15±0.11	7.77±0.38	8.04±0.29
Salinity	14.68 ± 8.23	6.49 ± 7.04	18.62 ± 9.07
TSS	15.00 ± 9.22	16.31 ± 20.17	11.00 ± 8.85
DO	7.08 ± 1.42	7.10 ± 1.29	7.19 ± 0.86

BOD	1.82 ± 0.86	2.09 ± 0.71	2.04 ± 1.22
NOrN	0.53 ± 0.22	0.62 ± 0.30	0.57 ± 0.37
NO3-N	2.74 ±1.17	3.79 ± 1.34	3.38 ±1.71
N114-N	2.63 ± 1.82	4.79 ± 1.86	3.46 ± 1.66
TN	28.01 ± 10.59	41.02 ± 10.94	69.75 ± 36.51
PO4-P	3.24 ± 1.15	11.44 ± 9.68	8.10 ± 2.81
TP	6.62 ± 2.92	14.49 ± 12.12	16.32 ± 10.93
Slat	3.54 ± 1.01	4.34 ± 1.11	6.53 ± 2.91
Chia (mg/m^3)	2.98 ± 2.03	4.25 ± 1.89	1.94 ± 0.59

TSS—Total suspended solids, TN—Total nitrogen, TP—Total phosphorus.

ganic load from the sewage of Paradip port, town ship as well as effluents from fertilizer industry situated upstream of the Mahanadi estuary [6]. The seasonal variations of BOD varies from 0.83 mg/l and 3.92 mg/l. BOD values highest in post monsoon due to the large amount of municipal sewage wastes, effluents from industries, Atharbanki creek carries all the sewage from Paradip township and industrial effluents from fertilizer industry directly into the estuary in this period.

The seasonal variations of nutrients concentrations during the sampling period ranged between 0.10 μmol/l l to 1.16 μmol/l NO_2, 1.32 μmol/l to 6.32 μmol/l NO_3, 0.63 μmol/l to 7.93 μmol/l NH_4, 17.13 μmol/l to 126.73μmol/l TN, 1.41 μmol/l to 28.97 μmol/l PO_4, 3.84 μmol/l to 36.98 μmol/l TP and 1.99 μmol/l to 10.55 μmol/l SiO_4. NO_2 always indicates the fresh input of organic load in to the water system [6]. Municipal sewage loads are accounted as the main source of organic matter for river and estuarine environment. NO_2 concentration was higher in post monsoon, which is due to the influx of municipal sewage from Paradip port and township. The NO_2 discharge through effluents during indicates that the concentration is diluted during post monsoon season. This phenomenon is also found in the Periyar river estuarine system [9]. The highest concentration NO_3 are found in post monsoon, due to the river water and agricultural runoff, Atharbanki creek carries all the sewage Paradip township and industrial effluents from fertilizer industry in to the Mahanadi estuary [17,18]. This phenomenon was observed in the Mahanadi river-estuarine system [6], the Godavari river-estuarine system [19], the Mandovi Zuari river-estuarine system [12] and Periyar river estuarine system [9]. The ammonia concentration is comparatively higher during post monsoon then the pre monsoon and summer. The source of ammonia is from localized anthropogenic input rather then River run-off. The Total nitrogen (TN) concentration is higher in summer, which may be due to the decay of organic debris, disintegration of industrial effluents and organic wastes. The highest concentration PO_4 and Total phosphorus recorded in post-monsoon of 2008-09 might be due to the effluent discharge from the near by Phosphatic industries. The spatial and seasonal variation total phosphorus high in summer, which may be that a higher value of organic from of phosphate is contributed through the municipal sewage of Paradip port town and effluents of Phosphatic industries. The seasonal variation show that silicate concentration is maximum during the summer season, probably on account of the input of more siliceous sediment gathered from its catchments area [1,20-22]. The above nutrients were higher in low tide than that of high tide. The minimum DO, BOD and other nutrient our present study were consider as polluted as the estuary, which is one of the most polluted estuaries east cost of India. Land runoff through the river from a large basin is the principal source of nutrient in the coastal water off Paradip, Bay of Bengal. The increase in industrial activities near Paradip found to have adverse impact on health of the Mahanadi river, estuary and its coastal environment. The growth of industrialization and urbanization is in the upstream and estuarine and coastal resources.

Chlorophyll-a concentration revealed wide seasonal variation. Chlorophyll-a varied from 1.35 to 6.92, 1.54 to 7.41 and 0.99 to 2.47 mg/m^3 for pre monsoon, post monsoon and summer seasons respectively. Relatively, higher values are also observed during post monsoon season reflects the higher phytoplankton productivity. The variation of Chlorophyll-a content during post monsoon is insignificant and a reverse condition is marked during pre-monsoon season. The chlorophyll-a concentration was observed higher during post monsoons.

STATISTICAL ANALYSIS

In recent years, various statistical procedures based on multivariate data from river-estuarine system have been used to formulate environmental classification, which helps for a better understanding of the chemical processes and difference in environmental variables, nutrient concentrations in the river-estuarine system. For better understanding of natural and anthropogenic fluxes responsible for characterization of water quality in Mahanadi estuarine system, researchers have carried out statistical analysis of the water quality study [6,23,24]. The present study includes the principal component analysis for three seasonal sets of data. pH shows positive significant correlation with salinity. Salinity shows negative significant correlation with NH_4 whereas DO shows negative significant correlation with BOD, and NO_3. BOD shows positive significant correlation with NO_2, NO_3 and NH_4. NO_2 has positive significant correlation with NH_4, PO_4. NH_4 shows positive significant correlation with PO_4 whereas TN produces positive significant correlation with TP and SiO_4 and PO_4 shows positive significant correlation with TP. The statistical analysis data with its seasons is represented graphically in figure 2.

CONCLUSIONS

A detailed seasonal variations physicochemical and nutrient content study of the estuarine water of Mahanadi over a period of two years brought out the following facts. The main source of pollutants is the residential one, generating both organic and inorganic wastes. Mahanadi estuary receives effluents from two fertilizer plants, from other small industries and domestic sewages from Paradip Township. Apart from these, fishing harbor activities at the estuary also affect the water quality. Atharbanki Creek is heavily loaded with the most of the municipal sewage of Paradip Township and effluents from industries such as Paradip Phosphate Limited (PPL) and passes it to Mahanadi estuary, which is due to influx of solid wastes from the Phosphatic fertilizer plants situated in the estuarine region. These wastes are allowed to join the sewers untreated bodies and create contaminations. During summer, reduced water volume and accelerated growth of microbes in higher temperature are responsible for higher degradation of organic matter, which eventually depleted the DO concentration. Domestic sewage appears to be the major source of pollutant in these water bodies. Results also indicate that the Mahanadi estuary with such overloaded situation cannot sustain any further sewage discharge. If proper alternative arrangements like sewage treatment before discharge are not made then the situation may be alarming to the inhabitants in the study area and to the downstream as well. Thus, adequate attempts have to be made to treat the waste water before discharging it to the water bodies to keep the marine environment healthy.

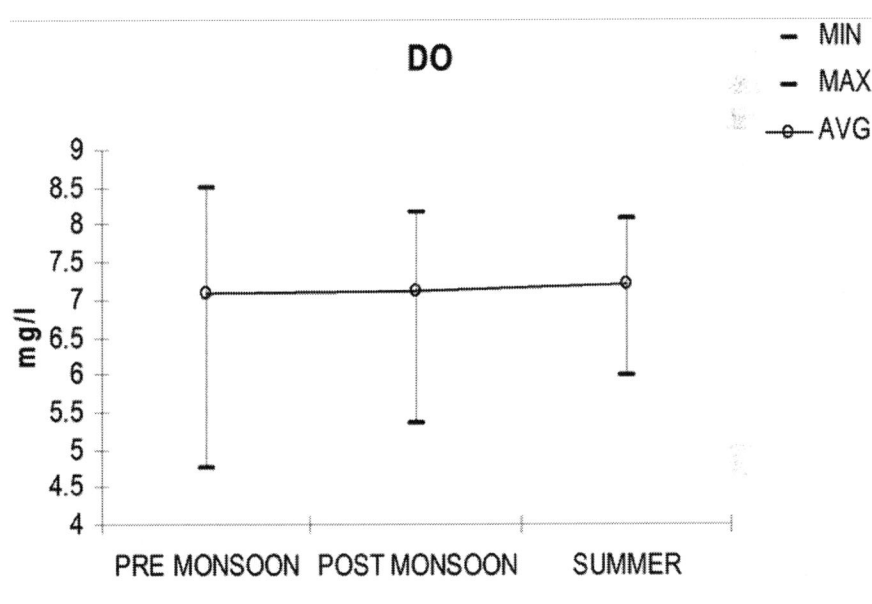

(a)

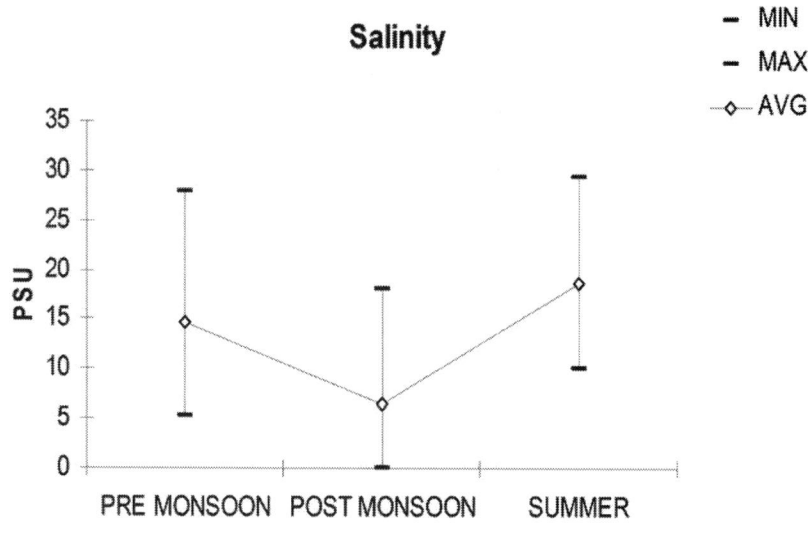

(b)

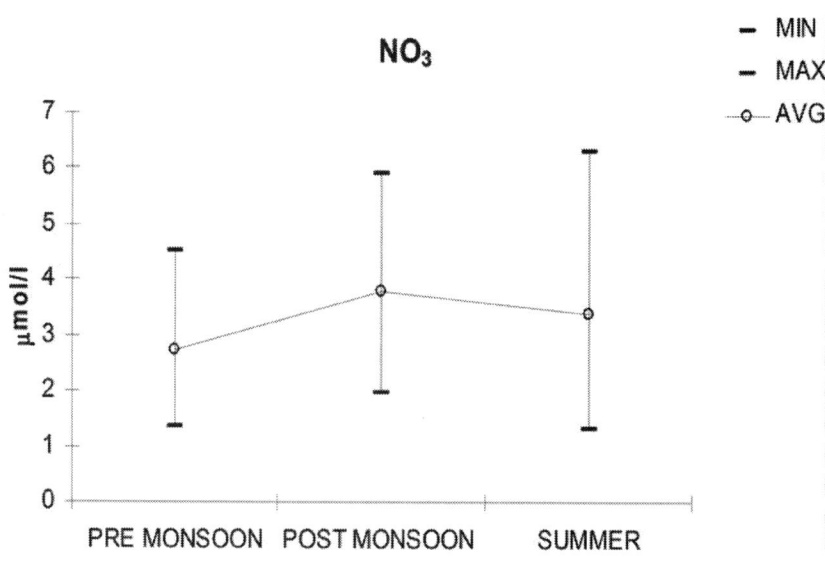

(c)

(d)

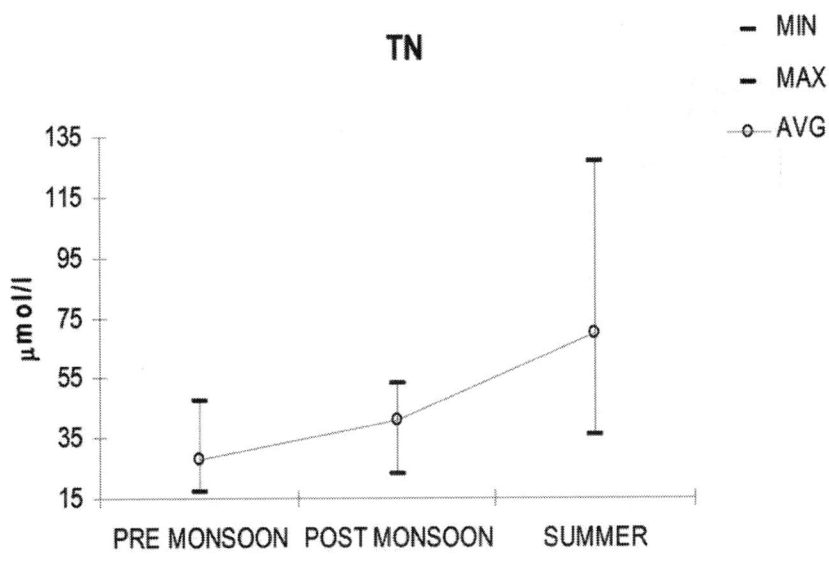

(e)

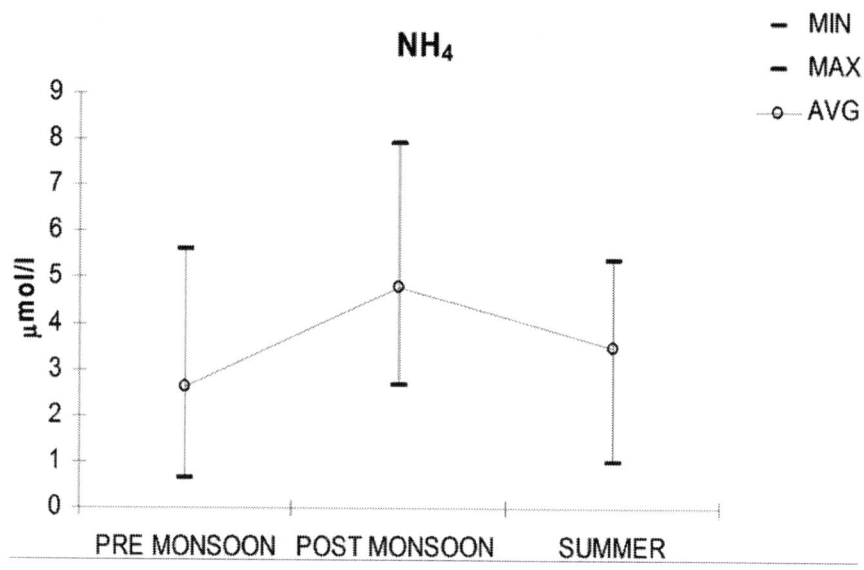

(f)

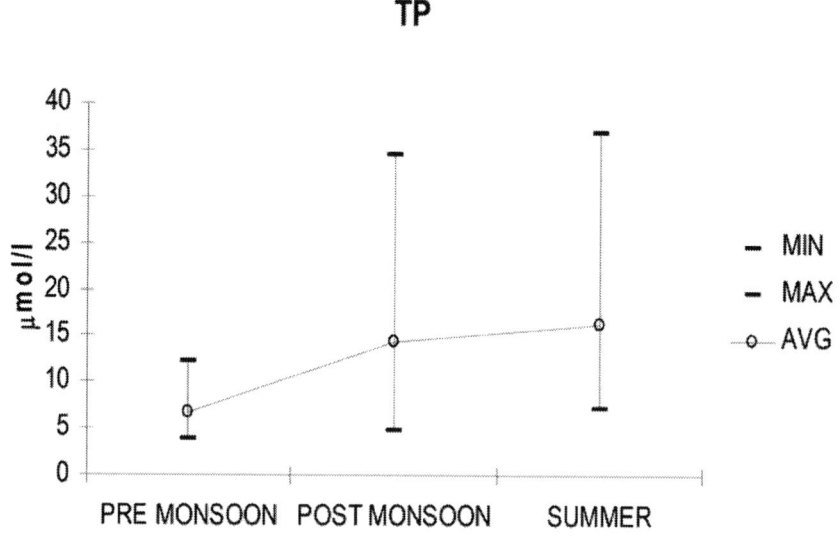

(g)

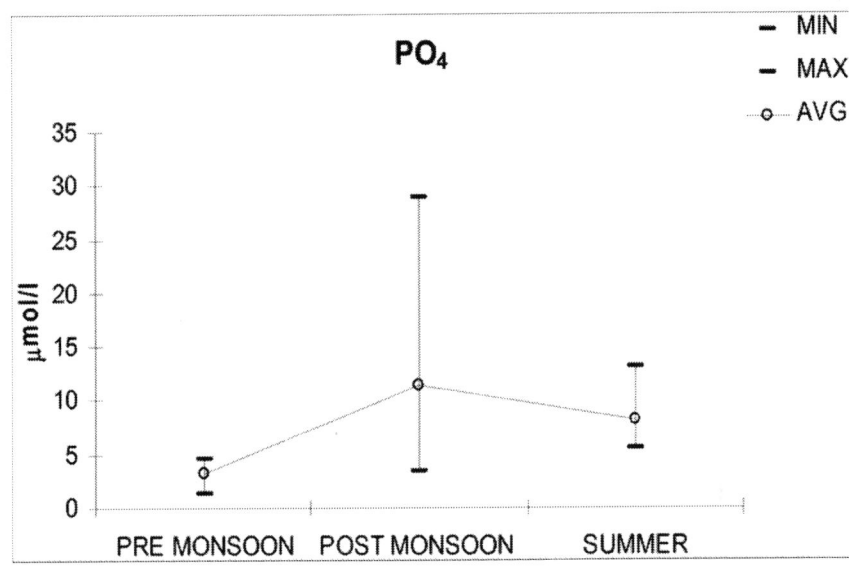

(h)

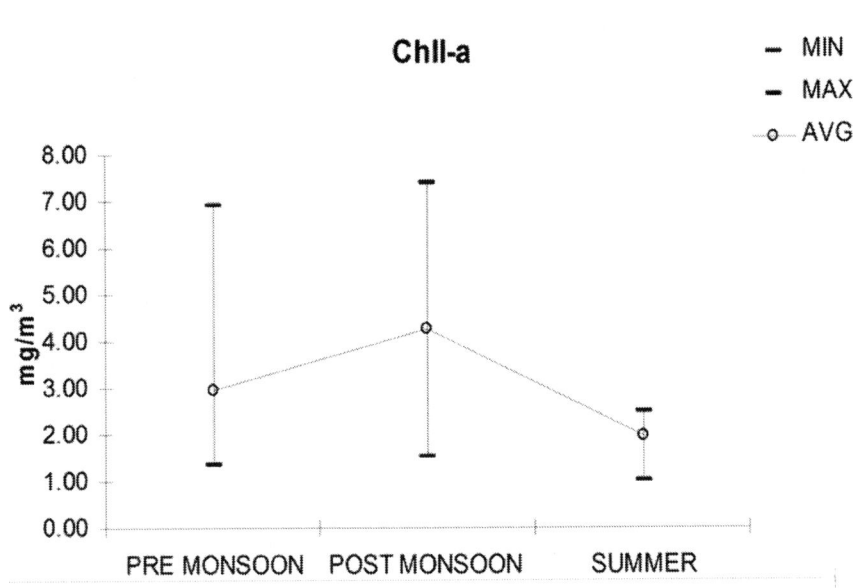

(i)

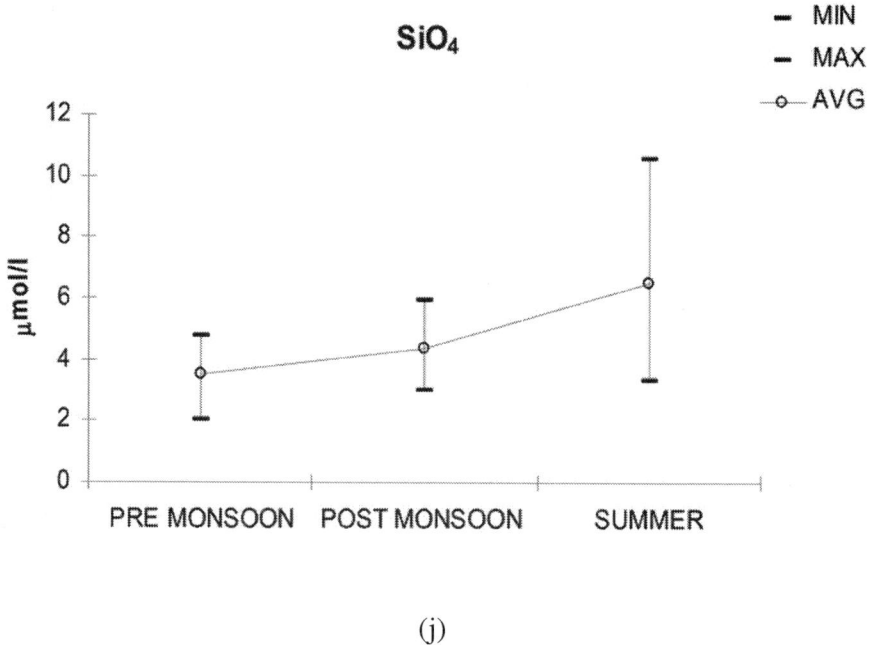

(j)

Figure 2: Seasonal variation of physicochemical parameters with standard deviation and its relation in variables of Mahanadi estuary in the years 2008-10.

ACKNOWLEDGEMENTS

The authors would like to acknowledge the encouragements from Director, Institute of Minerals and Materials Technology (CSIR), Bhubaneswar, Founder of KIIT University, Bhubaneswar for their continuous moral support and financial Support by MoES, Govt. of India and without which this innovative activities would not have been possible.

REFERENCES

1. R. Pai and M. P. M. Reddy, "Distribution of Nutrients of Malpe, South Kanara Coast," Indian Journal of Marine Science, Vol. 10, No. 4, 1981, pp. 322-326.

2.   G. J. Chakrapani and V. Subramanian, "Rate of Erosion and Sedimentation in the Mahanadi River Basin, India," Journal of Hydrology, Vol. 149, No. 1-4, 1993, pp. 39-48. doi:10.1016/0022-1694(93)90098-T

3.   B. B. Nayak, J. Das, U. C. Panda and B. C. Acharya, "Industrial Effluents and Muncipal Sewage Contamination of Mahanadi Estuarine Water, Orissa," Proceedings, Published Allied Publishers Pvt. Ltd., New Delhi, India, 2002, pp. 77-86.

4.   B. B. Nayak, B. N. Sahoo, B. C. Acharya and R. K. Sahoo, "Studies on Water and Sediment Qualities around Dhamara Estuary, Orissa," Vistas in Geological Research, Utkal University Spl. Publication in Geology, Vol. 2, 1997, pp. 243-250.

5.   I. Radhakrishna, "Saline Fresh Water Interface Structure in Mahanadi Delta Region, Orissa, India," Environmental Geology, Vol. 40, No. 3, 2001, pp. 369-380. doi:10.1007/s002540000182

6.   S. K. Sundaray, U. C. Panda, B. B. Nayak and D. Bhatta, "Behaviour and Distribution Pattern of Nutrients in RiverEstuarine Waters of Mahanadi, Orissa, India," Asian Journal of Water, Environment and Pollution, Vol. 2, No. 1, 2005, pp. 77-84.

7.   S. K. Sundaray, U. C. Panda, B. B. Nayak and D. Bhatta, "Multivariate Statistical Techniques for the Evaluation of Spatial and Temporal Variation in Water Quality of the Mahanadi River—Estuarine System (India)—A Case Study," Environmental Geochemistry and Health, Vol. 28, No. 4, 2006, pp. 317-330.

8.   K. O. Konhauser, M. A. Powell, W. S. Fife, F. J. Longstaffle and S. Tripathy, "Trace Element Geochemistry of River Sediment, Orissa State, India," Journal of Hydrology, Vol. 193, No. 1-4, 1997, pp. 258-269.

9.   K. Sarala Devi, V. N. Sankaranarayanan and P. Venugopal, "Distribution of Nutrients in the Periyar River Estuary," Indian Journal of Marine Sciences, Vol. 20, 1991, pp. 49- 54.

10.   G. J. Chakrapani and V. Subramanian, "Preliminary Studies on the Geochemistry of the Mahanadi Basin, India," Chemical Geology, Vol. 81, No. 3, 1990, pp. 241- 253. doi:10.1016/0009-2541(90)90118-Q

11.   M. Vega, R. Pardo, E. Barrado and L. Deban, "Assessment of Seasonal and Polluting Effects on the Quality of River Water by

Exploratory Data Analysis," Water Research, Vol. 32, No. 12, 1998, pp. 3581-3592. doi:10.1016/S0043-1354(98)00138-9

12. S. N Desouza, R. Sengupta, S. Sanzigiri and M. D. Rajgopal, "Studies on Nutrients of Mandovi and Zuari River Systems," Indian Journal of Marine Sciences, Vol. 10, 1981, pp. 314-321. doi:10.1007/s10653-005-9001-5

13. U. C. Panda, S. K. Sundaray, P. Rath, B. B. Nayak and D. Bhatta, "Application of Factor and Cluster Analysis for Characterization of River and Estuarine Water System— A Case Study, Mahanadi River (India)," Journal of Hydrology, Vol. 331, No. 3-4, 2006, pp. 434-445. doi:10.1016/j.jhydrol.2006.05.029

14. APHA, AWWA, WEF, "Standard Methods for the Examination of Water and Waste Water," 20th Edition, American Public Health Association, Washington DC, 1998.

15. K. Grasshoff, M. Ehrhardt and K. Bkremling, "Methods of Seawater Analysis," Wiley-VCH, Hoboken, 1999, pp. 159-226.

16. J. D. H. Strickland and T. K, Parsons "A Practical Hand Book of Seawater Analysis," Bulletin. Fisheries Research Board of Canada, Vol. 167, 1972, p. 172.

17. S. Upadhyay, "Physico-Chemical Characteristics of the Mahanadi Estuarine Eco-System, East Coast of India," Indian Journal of Marine Science, Vol. 17, 1988, pp. 19-23.

18. J. Das, S. N. Das and R. K. Sahoo, "Semidiurnal Variation of Some Physico-Chemical Parameters in the Mahanadi Estuary, East Coast of India," Indian Journal of Marine Science, Vol. 26, 1997, pp. 323-326.

19. A. K. Bhattacharya, A. Choudhury and A. Mitra "Seasonal Distribution of Nutrients and Its Biological Importance in Upper Stretch of Gangetic West Bengal," Indian Journal of Environment and Ecoplanning, Vol. 6, No. 3, 2002. pp. 421-424.

20. C. B. Sharma and N. C. Ghose, "Pollution of the River Ganga by Municipal Waste: A Case Study from Patna," Journal of the Geological Society of India, Vol. 30, No. 5, 1987, pp. 369-385.

21. A. K. Bhattacharya, A. Choudhury and A. Mitra "Seasonal Distribution of Nutrients and Its Biological Importance in Upper Stretch of Gangetic West Bengal," Indian Journal of Environment and Ecoplanning, Vol. 6, No. 3, 2002. pp. 421-424.

22. R. Gouda and R. C. Panigrahy, "Seasonal Distribution and Behaviour of Silicate in the Rushikulya Estuary, East Coast of India," Indian Journal of Marine Sciences, Vol. 21, No. 2, 1992, pp. 111-115.

23. P. K. Panigrahy, J. Das, S. N. Das and R. K. Sahoo, "Evaluation of the Influence of Various Physico-Chemical Parameters on Coastal Water Quality, around Orissa, by factor Analysis," Indian Journal of Marine Sciences, Vol. 28, No. 4, 1999, pp. 360-364.

24. S. K. Sundaray, U. C. Panda, B. B. Nayak and D. Bhatta, "Multivariate Statistical Techniques for the Evaluation of Spatial and Temporal Variation in Water Quality of the Mahanadi River—Estuarine System (India)—A Case Study," Environmental Geochemistry and Health, Vol. 28, No. 4, 2006, pp. 317-330.

A Simulation Study of Support Break-Off and Water Inrush during Mining under the High Confined and Thick Unconsolidated Aquifer

Yu Liu[1], Qimeng Liu[1], Zhouyang Jin[1], Liyong Cai[2], and Xueli Cui[1]

[1]School of Earth Science & Environmental Engineering, Anhui University of Science & Technology, Huainan, China
[2]Faculty of Engineering and Information Sciences, University of Wollongong, Wollongong, Australia

ABSTRACT

The thick Cenozoic unconsolidated aquifer is deposited under Sunan syncline core in Huaibei coalfield, the water yield property of unconsolidated bottom aquifer is strong and water pressure is high in some areas (up to 4 MPa in some areas). Water inrush accident often occurs during mining under unconsolidated aquifer, the biggest characteristic is abnormal mine pressure and support break-off during water inrush accident comparing with normal condition. In order to study mechanism of support break-off and water inrush during mining under the high confined thick unconsolidated aquifer, a simulation of similar material was designed. The experimental results indicated that, under normal condition, the compound breakage sequence of water-resisting key strata between coal seam and high confined thick unconsolidated aquifer is from top to bottom and the basic reason of synchronous fracture is the load of bottom key strata increased suddenly when the breakage of top key strata happened. Because of high confined thick unconsolidated aquifer, surface acts on the bottom key strata soil layer in the form of uniformly distributed load, which is the load-transfer mechanism of confined thick unconsolidated aquifer. Once the overlying key strata compound breaks, the height of unstable strata will reach far more than 30 meters and exceed support capability of current fully-mechanized mining supporter, which leads to support break-off accident during mining process under confined unconsolidated aquifer.

INTRODUCTION

Most of coalfield in North China are concealed coalfield covered by large-thick Cenozoic unconsolidated layers [1] . There is an aquifer in the bottom of unconsolidated layers consists of high permeability of unconsolidated sand and gravel where water yield property is good [2] [3] . The aquifer is directly hosted in the top of coal measures sandstone, commonly named as Fourth Aquifer or Bottom Aquifer. If the thickness of unconsolidated layers is great, the water pressure of the aquifer is normally high [1] [4] [5] .

In recent years, there are dozens of water inrush accidents happened during coal mining practice under high confined unconsolidated aquifer in China [6] [7] , which is a serious threat to coal mine safety production. There is always abnormally-increased roof pressure and even support break-off accidents happened accompanied with this kind of water inrush accident [8] . The roof pressure increased periodically with the advance of working face which shows that this kind of accident is not only associated with hydrogeology characteristics of high confined unconsolidated aquifer but also associated with structure types and the breakage characters of overburden strata [9] -[11] . The study of surrounding rock strata displacement and deformation is necessary technical foundation for studying coal mine water inrush mechanism and designing water control measures [8] [12] .

Qidong coal mine, affiliated to the Anhui Hengyuan Coal and Electricity Co., Ltd, is located in the north of Huainan-Huaibei Coal Mine field [6] [7] . There were many water inrush accidents occurred during mining process under high confined unconsolidated aquifer. For instance, there was a support break-off and water inrush accident in 3222-Working Face, the first coal mining face of Qidong Coal Mine after construction. The water inrush accident led to coal mine being submerged and 36.48 million RMB in direct economic loss and the indirect economic loss was almost 100 million RMB. After recovery production, similar water inrush accidents occurred in 3221, 7114, 6130 and 7130 working faces which had a severe impact on safe and high-efficient exploitation of coal mine.

Similar-material simulation experiment is a way to study the laws of nature by utilizing the similar features between object or phenomena which is based on the Similarity Theory [13] [14] . It is not only suitable in such research fields that hard to obtain research results by theory approach, but also an effective means to analyze and compare research results [15] . The model was made according to a certain proportion of the actual rock strata by similar material based on Similarity Theory. We excavated the model on the basis of the actual mining method, then observed the displacement, deformation and failure of upper rock strata and obtained relevant parameters. Thus, we can analyze and infer the displacement and deformation laws of surrounding rock as the working face advances [13] . According to three laws of similarity, the major requirement is geometric similarity, motion similarity and dynamic similarity when two systems are similar. Coal mining condition under high confined thick unconsolidated aquifer was modeled based on similar material simulation experiment. We also obtained stress distribution law and roof displacement characteristics of surrounding rock strata, which provided technical support to further study of support break-off water inrush mechanism during mining under high confined thick unconsolidated aquifer.

DESIGN OF SIMULATION EXPERIMENT

Experiment Objective

The experiment objective is to study load-transfer mechanism of confined unconsolidated aquifer and laws of key strata compound breakage; compare and analyze the influence of confined unconsolidated aquifer on sequence of key strata breakage and explore the reason that support break-off water inrush often occurs during mining under confined aquifer.

Model Prototype

Research area topography is flat and surface elevation is +21 m which located in the middle part of Huaibei Plain in Anhui Province. Hui river is a seasonal river and is the tributary of Huaihe river which flows through south part of coalfield. Sequence stratigraphy from top to bottom in the research area is Quaternary (Q), Neogene (N), Palaeogene (E), Cretaceous (K), Jurassic (J), Triassic (T), Permain (P), Carboniferous (C_{1+2}) and Ordovician (O_{1+2}). Carboniferous-permian is the main coal-bearing strata of Qidong coal mine. Coal-bearing strata are covered with Cenozoic unconsolidated layers which belong to concealed coalfield. Geological structure is comparatively simple in Qidong coal mine. There are three faults and monoclinal structure in research area. Occurrence of coal seam and rock stratum are different from West to East. The inclination of rock stratum turns from NNW trending to NNE trending. Fracture water of main coal seams' roof and floor sandstone is the direct source of water during mining. Main recharge of sandstone fracture water is infiltrate of unconsolidated aquifer in the bottom of Cenozoic. Thickness of Cenozoic unconsolidated layers is 350 - 375 m varying with ancient landform. Under the influence of neotectonic fault control and basin landform, thickness of Cenozoic unconsolidated layers become bigger and bigger from NE trending to SW trending. Lithology of research area consists of clay, sandy clay, clayey sand, silt, fine sand, medium sand and gravel, etc. The hydrogeological structure consists of multilayer of aquifers and water-resisting layers. The lithology and mechanical parameters of rock strata are shown in Table 1 [16] . According to the objective of experiment, the model prototype was designed, shown as Figure 1.

Calculation of Similar Condition

Geometrical scale: $\alpha_l = 0.01$,

Poisson's ratio scale:, $C_\mu = \mu_m / \mu_p = 1$,

Volume weight scale:, $\alpha_\gamma = \gamma_m / \gamma_p = 0.6$,

Stiffness scale:, $C_E = E_m / E_p = 0.6$,

Stress scale:, $\alpha_R = \alpha_m / \alpha_p = C_E = 0.6$, and

Model time scale: $\alpha_t = 0.1$.

where,

γ_p—Volume weight of actual rock strata, γ_m—Volume weight of rock strata in the model;

γ_m—Poisson's ratio of actual rock strata, μ_m—Poisson's ratio of rock strata in the model;

E_p—Stiffness of actual rock strata, E_m—Stiffness of rock strata in the model;

α_p—Stress of actual rock strata, α_m—Stress of rock strata in the model.

Experimental Equipment

The experimental equipment consists of test bench, confined unconsolidated aquifer (bags, filled with water and sand) and water pressure control device, as shown in Figure 2.

Table 1: The lithology and mechanical parameters of rock strata in the model prototype

Lithology	Volume weight/ N·m−3	Rock strata thickness/m	Elasticity modulus/ GPa	Poisson's ratio	Cohesion/ MPa	Internal friction angle/(°)	Strength of extension/ MPa	Remark
Key strata 2	25,000	8	60	0.4	40	40	7.0	Thickness of stratification is 2 m
Soft rock 2	25,000	8	10	0.3	30	30	1.0	
Key strata 1	25,000	8	60	0.4	40	40	7.0	
Soft rock 1	25,000	8	10	0.3	30	30	1.0	
Coal seam	14,000	3	10	0.2	10	15	1.0	
Floor	25,000	5	60	0.3	30	40	7.0	

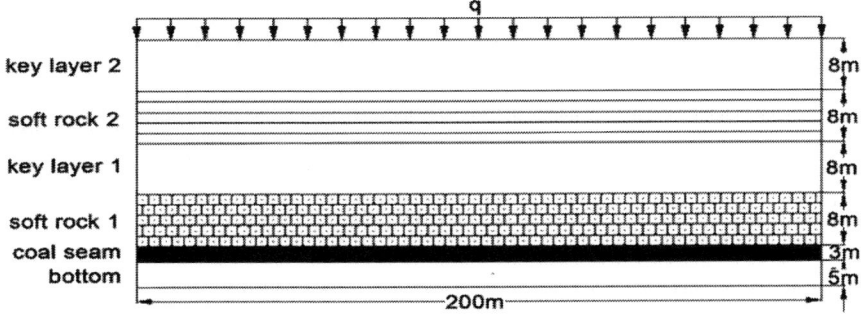

Figure 1: Similar simulation prototype graph.

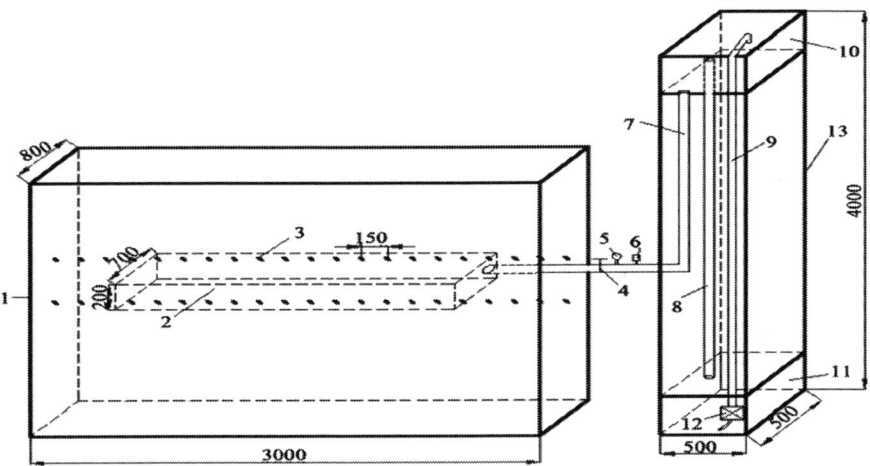

Simulation platform, 2-Bag be filled with water and sand (confined aquifer), 3-Pressure Sen-sor, 4-Valve, 5-Pressure gauge, 6-Pressure Sensors, 7-Injection Hose; 8-Overflow Hose; 9-Pipes up, 10-Upper tank, 11-Pipes down, 12-Pumps, 13-Tank lifting frame

Figure 2: Test device of unconsolidated confined aquifer load transfer function (Unit: mm).

Requirements of model equipment:

- The major measuring instrument is pressure transducer and the sensitivity of pressure transducer and reliability of data gathering must be guaranteed. The sensitivity of pressure sensor must meet requirement of 1% and maximum value of 0.1 MPa.

- The equipment need to be a high maneuverability. Water supply of pump has to reach 100 L/min.

- Pressure transducers are put in the top and bottom surface of confined unconsolidated aquifer. The space between two pressure transducers is 150 mm.

- Injection hose, overflow hose and upper hose must be soft and total length should surpass the maximum height of tank lifting frame.

Program control static resistance strain gauge is used to transform pressure obtained by pressure transducers [17] , computer is used to collect and save data, digital camera is used to shoot mining process of model and typical failure characteristics of model.

Experimental Material

Gypsum and lime were taken as cementing material in the experiment and gypsum is major. Sand, calcium carbonate, pulverized coal and water are the main material of coal seam and strata simulation. Ordinary river sand (particle size less than 1.5 mm) is aggregate material and mica powder is layered material. Determining the amount of experimental material is based on type and volume of model material. Material ratio and amount of each rock strata are shown in Table 2.

Experimental Schemes

Determine the following three simulation project based on experiment objective.

The First Simulation Project

The first simulation project is used to study laws of key strata compound breakage under the uniform load q. We load by iron block and wooden tablet to assure the uniform load act on model. The mass of each iron block is 3.23 kg and the contact area with model is 0.02 m^2. Thus the uniform load q is 1.27×10^4 Pa. The model simulation equipment based on similar simulation ratio is shown in Figure 3.

Table 2: Material ratio and amount of each rock strata in simulation

Rock strata ID	Lithology	Thickness/ cm	Sand/ kg	Gypsum/ kg	Calcium carbonate/ kg	Water/ kg	Pulverized coal/kg
1	Coal seam floor	5	35.0	2.2	2.8	5.0	
2	Coal seam	3	22.0	0.4	1.6	3.0	
3	Soft rock 1	10	66.7	4.0	9.3	10.0	1.5
4	Key strata 1	10	64.0	4.8	11.2	10.0	
5	Soft rock 2	10	66.7	4.0	9.3	10.0	1.5
6	Key strata 2	20	128.0	9.6	22.4	20.0	

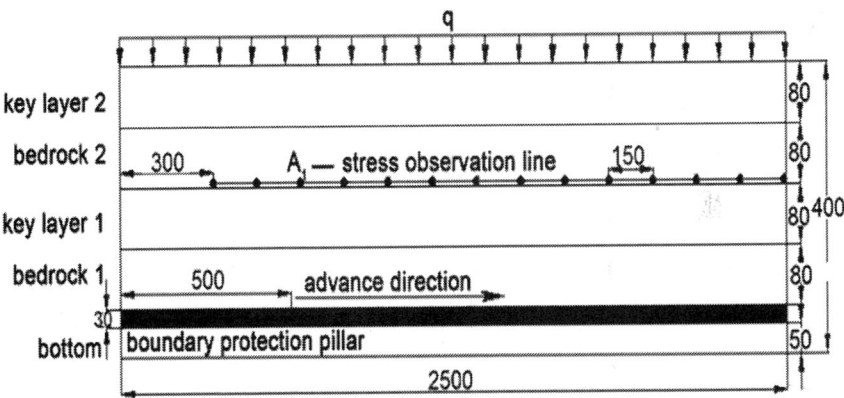

Figure 3: Schematic diagram of the test apparatus in Scenario 1 (Unit: mm).

Technical points:

- Set a 500 mm protection pillar at mining boundary to eliminate boundary effect;
- Put a stress observation line on the top surface of key strata 1 to observe load distribution and changing rule of stress. The particular arrangement of the pressure sensor above the observation line is shown in Figure 3.

The Second Simulation Project

The second simulation project is used to study load-transfer mechanism of confined unconsolidated aquifer. Thickness of surface soil is determined by uniform load in first simulation project, confined unconsolidated aquifer is put in the middle of surface soil and key strata 2. Transform the uniform load into surface soil with equivalent thickness on the basis of first simulation project. The volume weight of surface soil is 16 kN·m⁻³; equivalent thickness is 80 m based on geometrical scale. Surface soil consists of sand, gypsum, calcium carbonate and water which were divided into 50 layers successively. Aggregate material of confined unconsolidated aquifer is composed of stones with comparatively bigger size particles. Constant water pressure of confined aquifer can be controlled by the height of water surface. Schematic diagram and photograph of model simulation equipment are shown in Figure 4 and Figure 5 respectively.

   Technical points:

- Keep water pressure of confined aquifer steady and the pressure head of confined should match to the thickness of surface soil.

- Confined unconsolidated aquifer consists of aggregate material and water in the simulation, aggregate material is stones which particles with comparatively big size.

- Put three stress observation lines in the area between key strata and confined unconsolidated aquifer to observe the relationship between stress distribution of top and bottom surface of aquifer by changing mining width.

The Third Simulation Project

The third simulation project is used to study breakage sequence of two key strata without influence of confined unconsolidated aquifer. To ignore the effect of load transfer, we put the surface soil on the key strata 2 directly. Thus, breakage sequence of each key stratum on that situation can be obtained. Schematic diagram of simulation equipment are shown in Figure 6.

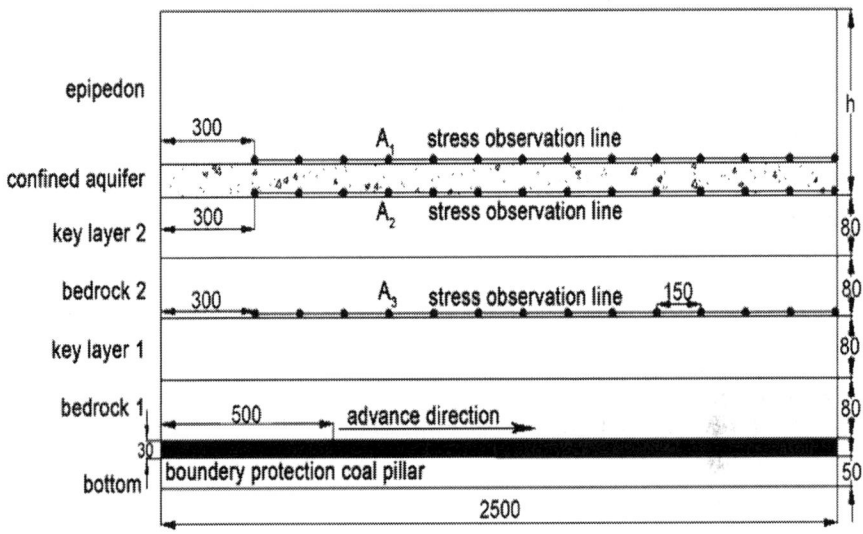

Figure 4: Schematic diagram of the test apparatus in Scenario 2 (Unit: mm).

Figure 5: Scheme 2 similar model objective graph.

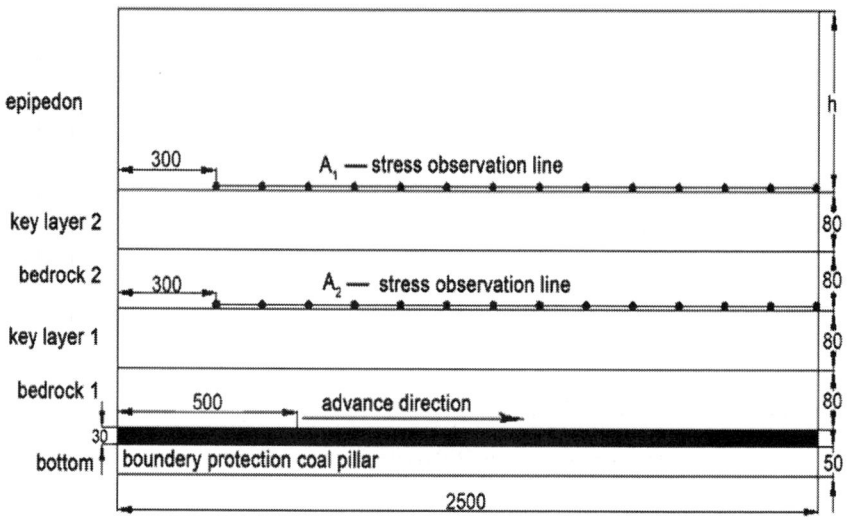

Figure 6: Schematic diagram of the test apparatus in Scenario 3 (Unit: mm).

PROCESS AND RESULT OF EXPERI-MENT

Process and Result of the First Simulation Project

Key strata 1 and 2 fractured at the same time when the mining width reached 85 cm in the model. Fracture surface can be seen obviously (Figure 7). Data collection situation is different from each other with different mining widths (Table 3). Figure 8 describes the relationship between stress and data collection time at observation point, 70 cm away from the mining boundary.

Stress changes at measure point can be divided into the following four stages depending on mining width (Figure 8).

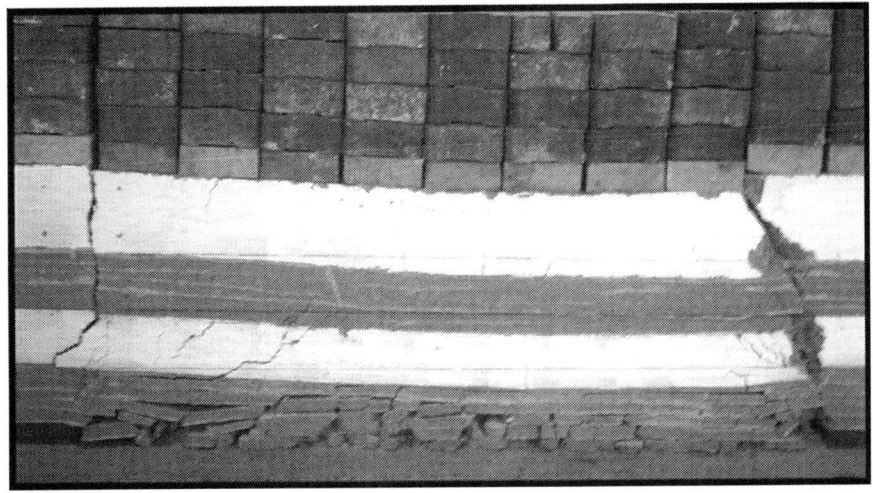

Figure 7: Key layer synchronization breaking in similar simulation.

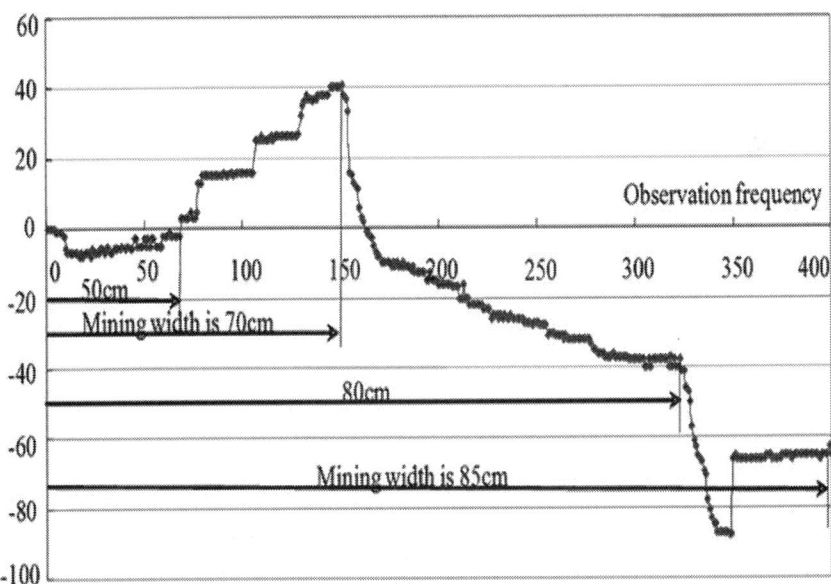

Figure 8: Stress changes at observation point, 70 cm away from the mining boundary.

Table 3: Distribute of data collection number with different mining width

Mining width/m	10	20	25	30	35	40	45
Data collection number	0 - 5	6 - 10	11 - 12	13 - 34	35 - 47	48 - 59	60 - 68
Mining width/m	50	55	60	70	80	85	
Data collection number	69 - 76	77 - 106	107 - 128	129 - 154	155 - 331	332 - 400	

The first stage: as mining width is between 0 cm and 50 cm, stress at measure point remains the same, belongs to primary rock stress [18].

The second stage: as mining width is between 50 cm and 70 cm, stress at measure point along with the mining width increases gradually, belongs to stress increase region.

The third stage: as mining width is between 70 cm and 80 cm, stress at measure point decreases gradually, belongs to stress decrease region.

The fourth stage: mining width is 85 cm, stress at measure point has a leap-growth based on the third stage.

Two key strata synchronous fracture happened because of roofing breaking interval in the fourth stage, the stress leap-growth at measure point is also the result of key strata breakage. From the aspect of macro behavior and variation of stresses at the moment of key strata breakage, mechanism of key strata compound breakage is top key strata break first, the reason of synchronous fracture is the load of bottom key strata increased suddenly as the breakage of top key strata.

Process and Result of the Second Simulation Project

Key strata 1 and 2 fractured at the same time when the mining width reached 90 cm in the model (Figure 9). Number of data collection is different with different mining width (as shown in Table 4). As Figure 10 shown that the relationship between stress and number of data collection at measure point of 95 cm to the mining boundary.

Figure 11 and Figure 12 showed the relationship between stress and number of data collection respectively at two measure points 70 cm to the mining boundary, at the top and bottom of confined unconsolidated aquifer.

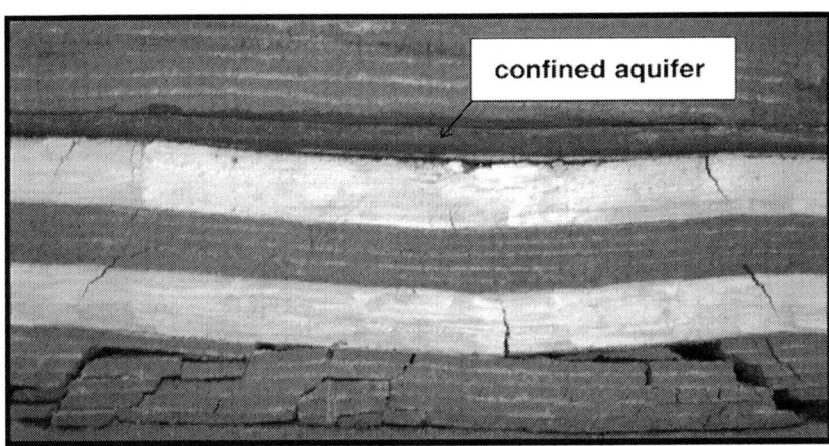

Figure 9: Key layer synchronization breaking in similar simulation.

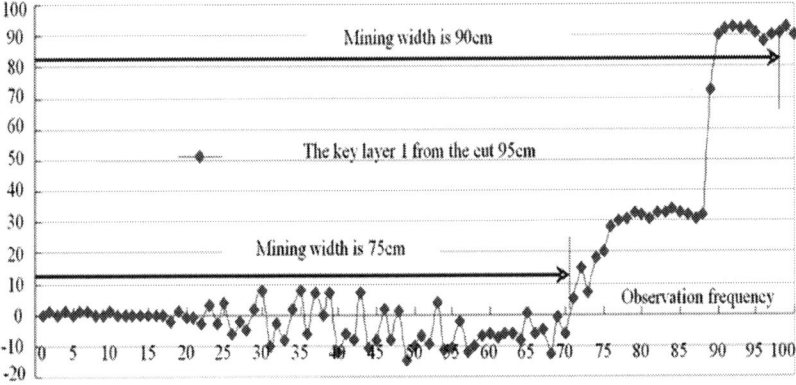

Figure 10: Stress changes at observation point, 95 cm away from the mining boundary.

Table 4: Distribution of data collection number with different mining width

Mining width/m	10	20	25	30	35	40	45
Data collection number	1 - 18	19 - 21	22 - 26	27 - 31	32 - 36	37 - 40	41 - 48
Mining width/m	55	60	70	75	80	90	
Data collection number	49 - 53	54 - 62	63 - 67	68 - 73	74 - 87	88 - 98	

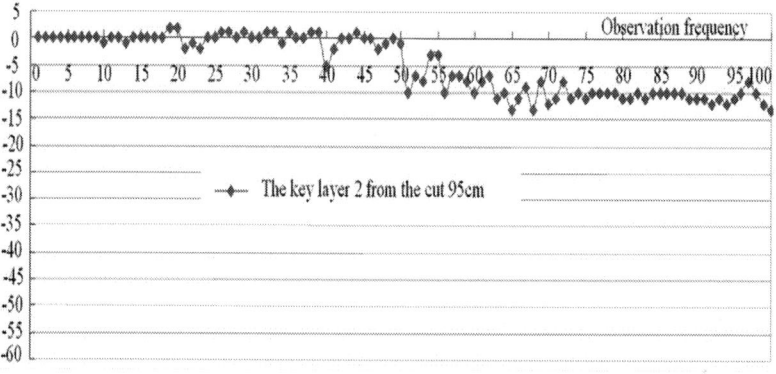

Figure 11: Stress changes at observation point, 70 cm away from the mining boundary.

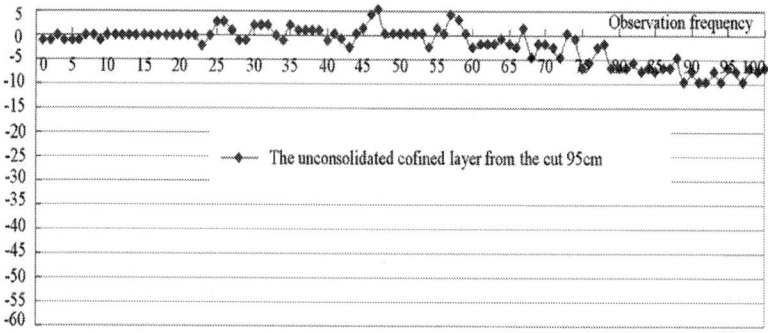

Figure 12: Stress change of observation points 70 cm away from mining boundary in unconsolidated confined aquifer.

Stress at the measure point 95 cm to the mining boundary can be divided into the following two stages depending on mining width.

The first stage: mining width is between 0 cm and 70 cm, belonging to primary rock stress.

The second stage: as mining width is between 70 cm and 90 cm, stress at measure point increases gradually and has a leap-growth at 90 cm.

Breakage of key strata 2 resulted in load increases of key strata 1. According to primary roofing breaking interval of two key strata and relationship between stress and mining width at measure point on key strata 1, breakage of key strata belongs to compound breakage as well in Scenario 2.

With the influence of confined unconsolidated aquifer, surface soil and key strata stress are substantially free from the impact of mining width. There is no significant stress reduction or increase. Stress is a constant value.

The second simulation showed that with the existence of confined unconsolidated aquifer between surface soil and key strata, laws of key strata breakage are the same with the influence of uniform load. The type of breakage is still compound breakage and there is no obvious stress increase or decrease. Thus, we can conclude that with the existence of confined unconsolidated aquifer, load-transfer mechanism of confined unconsolidated aquifer is surface soil act as uniform load on bottom rock strata all the time.

Process and Result of the Third Simulation Project

Key strata broke suddenly when the mining width reached 80 cm and surface soil and key strata 2 had no obvious changes. Key strata 2 broken and there was caving arch in surface soil when the mining width was up to 110 cm. Distribution of data collection number with different mining widths is shown in Table 5.

Figure 13 shows the relationship between stress and mining width at measure point of 60 cm to the mining boundary. Figure 14 shows the relationship between stress and mining width at measure point of 75 cm to the open-off cut. Figure 15 shows the stress distribution of each measure points on the key strata 2 when mining width is 80 cm, stress concentration and stress release emerged.

Figure 13 shows that relationship between stress and mining width can be divided into the following five stages:

Table 5: Distribution of data collection number with different mining width

Mining width/m	10	20	25	30	35	40	50
Data collection number	1 - 22	23 - 51	52 - 73	74 - 113	114 - 148	149 - 170	171 - 229
Mining width/m	60	65	75	80	90	100	110
Data collection number	230 - 278	279 - 356	357 - 376	377 - 419	420 - 451	452 - 469	470 - 509

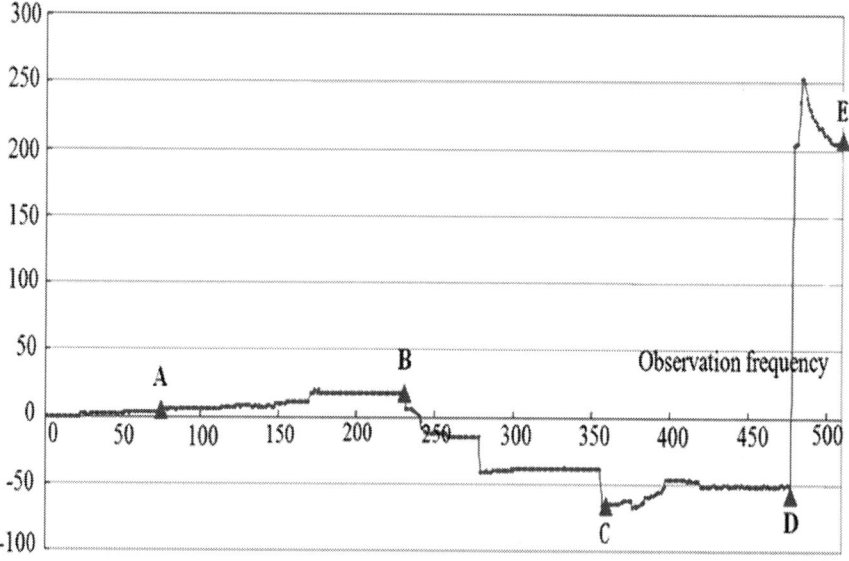

Figure 13: Stress change of observation points 60 cm away from mining boundary.

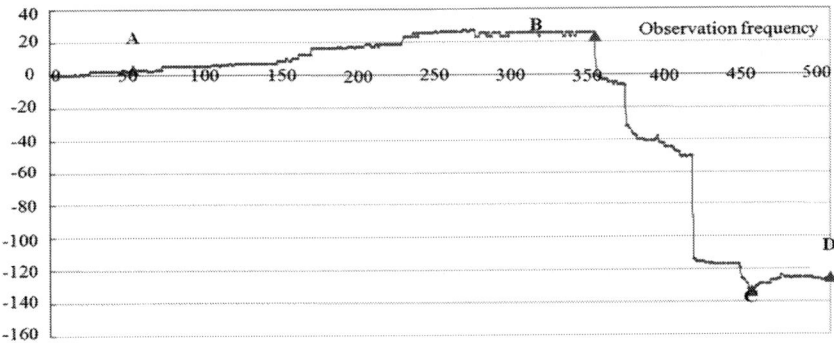

Figure 14: Stress change of observation points 75 cm away from mining boundary.

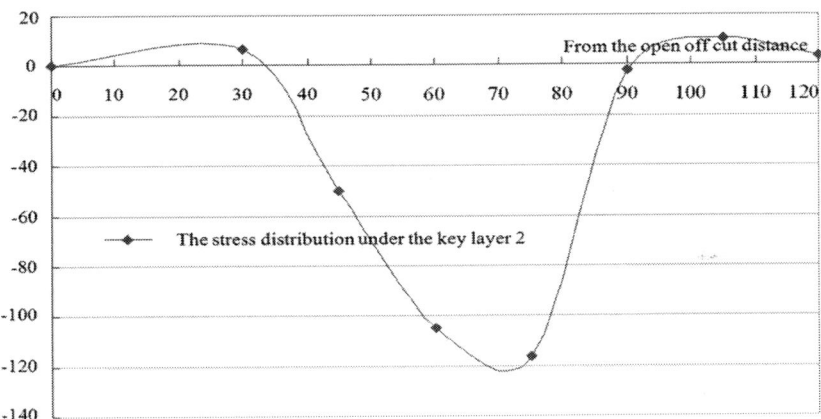

Figure 15: Stress distribution of observation points in key layer 2 when mining width is 80 cm.

O - A stage: mining width is 0 cm to 30 cm, belongs to primary rock stress,

A - B stage: mining width is 35 cm to 60 cm, stress increases,

B - C stage: mining width is 65 cm to 75 cm, stress decreases,

C - D stage: mining width is 80 cm to 105 cm, stress remains the same,

D - E stage: mining width is 110 cm, stress has a leap-growth and then decreases.

Unchanged remaining stress in C - D stage shows the abscission layer emerged between key strata 1 and key strata 2. All the weight of top layer was supported by key strata 2. According to primary roof breaking interval of key strata 2, key strata 2 breaking in D - E stage led to observed stress increase.

Compared with Figure 13, there was no C - D stage in Figure 8, which shows that stress distribution on key strata 1 between layered successive breakage and compound breakage are different.

Figure 14 shows that relationship between stress and mining width can be divided into four stages. Because that surface soil are unconsolidated, soil in caving arch keep synchronous movement with key strata 2 led to has no stress leap-growth during mining. Figure 15 shows the stress concentration and stress release emerged obviously on key strata 2 and load is following non-uniform distribution.

Comparing Figure 12 and Figure 14, stress decreased obviously without influence of confined unconsolidated aquifer. Mainly because of arching phenomenon emerged with the direct action of surface soil led the load act on key strata 2 far more less than the load induced by all weight of surface soil.

Key strata compound breakage was easy to emerge with the influence of confined unconsolidated aquifer. Key strata breakage type transform from successive breakage to compound breakage without influence of confined unconsolidated aquifer. Main reason is increasing mining width, arching phenomenon led the load act on key strata equivalent to the weight of soil in arch which far more less than the all soil.

With the influence of confined unconsolidated aquifer, the load act on key strata 2 remain the same along with mining step increasing. Relatively large load is the key factor led the synchronous compound breakage of key strata 1 and key strata 2 (Figure 16). Be different with above situation, without the influence of confined unconsolidated aquifer, the load act on key strata 2 decreased along with mining step increasing is the key factor led the unsynchronized of key strata. All above shows that load-transfer mechanism of confined unconsolidated aquifer is the direct reason of key strata breakage sequence transformation.

Key strata 2 and nearby rock strata are the load layer of key strata 1 when the key strata of surrounding rock compound breakage. Load on voussoir [9] beam structure consisting of breaking block increased obviously. Support break-off accident happened when retaining resistance of support less than the weight of unstabilized rock pillar. Retaining resistance of current fully mechanized mining supporter is about 500 KN and can afford about 30 m unstabilized rock pillar. Mining under confined unconsolidated aquifer, as far as key strata of surrounding rock compound breakage, height of unstabilized rock pillar is far more than 30 m and support break- off accident will happen inevitably.

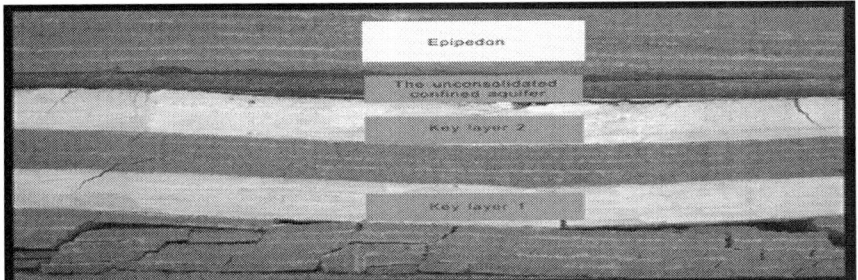

(a)

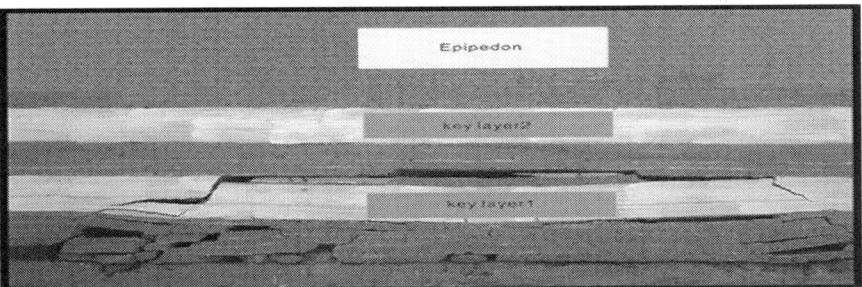

(b)

Figure 16: Test result of the unconsolidated confined aquifer influencing compound and fracture of key layer. (a) Key layer synchronization fracture in the condition of the unconsolidated confined aquifer; (b) Key layer fracture step by step in the condition of the non-unconsolidated confined aquifer.

CONCLUSIONS

Normally, the sequence of key strata break is the above key strata breaking off firstly, and load of the lower key strata suddenly increases in an instant of the above breaking, which is the basic reason of synchronous breaking with the above key strata.

As the unconsolidated confined aquifer exists between topsoil and the lower key strata, the load is generated by the topsoil all on the lower key strata, which is due to the influence of the unconsolidated confined aquifer. Topsoil acting on the lower key strata through the uniform load is the load transfer mechanism of loosely confined aquifer.

When the complex breaking key strata is in non-unconsolidated confined aquifer under the unconsolidated confined aquifer condition, the key strata turned layered successive breaking from original synchronous breaking and the load acting on the key strata is just the result of rock gravity within the height of caving arch, which is much smaller than the load generated by the thickness of topsoil.

As the key strata of overburden rock occurs compound breakage, there will be the instability of masonry beam structure and when the bracket support resistance is less than the weight of unstable rock pillars, it will cause the support crushing accident. Mining was done under the unconsolidated confined aquifer. Once compound breakage happened in the key strata and the bracket could not support the pressure of the instability of rock, mining crushing accidents will happen.

ACKNOWLEDGEMENTS

This study was financially supported by the National Natural Science Foundation of China (No. 41472235) and Anhui Provincial Natural Science Research Project in Colleges and Universities (KJ2014ZD11). The authors also thank the anonymous reviewers and editors for their constructive comments and suggestions on improving this manuscript.

REFERENCES

1. Xu, J.L., Wang, X.Z., Liu, W.T., et al. (2009) Effects of Primary Key Stratum Location on Height of Water Flowing Fracture Zone. Chinese Journal of Rock Mechanics and Engineering, 28, 380-385.

2. Hang, Y., Zhang, G.L. and Yang, G.Y. (2009) Numerical Simulation of Dewatering Thick Unconsolidated Aquifers for Safety of Underground Coal Mining. Mining Science and Technology (China), 19, 312-316. http://dx.doi.org/10.1016/S1674-5264(09)60058-2

3. Paradis, D. and Lefebvre, R. (2013) Single-Well Interference Slug Tests to Assess the Vertical Hydraulic Conductivity of Unconsolidated Aquifers. Journal of Hydrology, 478, 102-118. http://dx.doi.org/10.1016/j.jhydrol.2012.11.047

4. Sui, W.T., Cai, G.T. and Dong, Q.H. (2007) Experimental Research on Critical Percolation Gradient of Quicksand across Overburden Fissures Due to Coal Mining Near Unconsolidated Soil Layers. Chinese Journal of Rock Mechanics and Engineering, 26, 2084-2091.

5. Xu, J.L., Cai, D. and Fu, K.L. (2007) Mechanism of Supports Crushing Accident and Its Preventive Measures during Coal Mining near Unconsolidated Confined Aquifer. Journal of China Coal Society, 32, 1239-1243.

6. Tan, S.Y. and Wu, J.S. (2006) Cause Analysis of Water Bursting in $7_1 14$ Mining Face of 7_1 Coal Seam in Qidong Colliery. Journal of Coal Mining Technology, 11, 64-67.

7. Xiong, X.Y. and Li, J.B. (2004) A Case Study of Support Break-Off at 1402(3) Fully Mechanized Mining Face. Coal Geology of China, 16, 34-37.

8. State Coal Industry (1984) Mine Hydro-Geological Point of Order. Coal Industry Press, Beijing. (In Chinese)

9. Xu, J.L. and Qian, M.G. (2000) Method to Distinguish Key Strata in Overburden Strata. Journal of China University of Mining and Technology, 29, 463-467.

10. Lu, H.F., Yuan, B.Y. and Wang, L. (2011) Rock Parameters Inversion for Estimating the Maximum Heights of Two Failure

Zones in Overburden Strata of a Coal Seam. Mining Science and Technology (China), 21, 41-47.

11. Li, J.K., Wang, J.A. and Cui, S.H. (2005) Study on Pump Excavation Deformation and Fracture with Complex Stress under Deep Mining and High Pressure. Ground Pressure and Strata Control, 22, 12-13.

12. Xiao, T.Q., Wang, X.Y. and Zhang, Z.G. (2014) Stability Control of Surrounding Rocks for a Coal Roadway in a Deep Tectonic Region. International Journal of Mining Science and Technology, 24, 171-176. http://dx.doi.org/10.1016/j.ijmst.2014.01.005

13. Guo, G.L., Zha, J.F., Miao, X.X., Wang, Q. and Zhang, X.N. (2009) Similar Material and Numerical Simulation of Strata Movement Laws with Long Wall Fully Mechanized Gangue Backfilling. Procedia Earth and Planetary Science, 1, 1089-1094. http://dx.doi.org/10.1016/j.proeps.2009.09.167

14. Lu, A.H., Mao, X.B. and Liu, H.S. (2008) Physical Simulation of Rock Burst Induced by Stress Waves. Journal of China University of Mining and Technology, 18, 401-405.http://dx.doi.org/10.1016/S1006-1266(08)60084-X

15. Li, Y.J., et al. (2014) A Physical and Numerical Investigation of the Failure Mechanism of Weak Rocks Surrounding Tunnels. Computers and Geotechnics, 61, 292-307.http://dx.doi.org/10.1016/j.compgeo.2014.05.017

16. Bieniawski, Z.T. (1989) Rock Mass Classifications in Rock Engineering. John Wiley & Sons, Inc, Hoboken.

17. Xiao, T.Q., et al. (2011) Characteristics of Stress Distribution in Floor Strata and Control of Roadway Stability under Coal Pillars. Mining Science and Technology (China), 21, 243-247. http://dx.doi.org/10.1016/j.mstc.2011.02.016

18. Liu, C.R. (2011) Distribution Laws of In-Situ Stress in Deep Underground Coal Mines. Procedia Engineering, 26, 909- 917. http://dx.doi.org/10.1016/j.proeng.2011.11.2255

Production of Natural Coagulant from Moringa Oleifera Seed for Application in Treatment of Low Turbidity Water

Md Zahangir Alam and Mohd Ramlan M. Salleh

Biotechnology Engineering Department, Faculty of Engineering, International Islamic University Malaysia, Kuala Lumpur, Malaysia

ABSTRACT

This study focused on developing an efficient and cost effective processing technique for Moringa oleifera seeds to produce natural coagulant for use in drinking water treatment. The produced natural coagulant can be used as an alternative to aluminum sulphate and other coagulants and used worldwide for water treatment. This study investigates processing Moringa oleifera seeds to concentrate the bio-active constituents which have coagulation activity. Moringa oleifera seeds were processed for oil extraction using electro thermal soxhlet. Isolation and purification of bio-active constituents using chromatography technique were used to determine the molecular weight of the bio-active constituents. The molecular weight of bio-active constituents found to be in a low molecular weight range of between 1000 – 6500 Dalton. The proposed method to isolate and purify the bio-active constituents was the cross flow filtration method, which produced the natural coagulant with very simple technique (oil extraction; salt extraction; and microfiltration through 0.45 µm). The turbidity removal was up to 96.23 % using 0.4 mg/L of processed Moringa oleifera seeds to treat low initial turbidity river water between 34-36 Nephelometric Turbidity Units (NTU) without any additives. The microfiltration method is considered to be a practical method which needs no chemicals to be added compared to other researchers proposed methods. The natural coagulant produced was used with low dosages to get high turbidity removal which considered to be a breakthrough in this study and recommended to be scaled up for industry level. The product is commercially valuable at the same time it is minimizing the cost of water treatment.

INTRODUCTION

Developing countries are facing potable water supply problems because of inadequate financial resources. The cost of water treatment is increasing and the quality of river water is not stable due to a suspended and colloidal particle load caused by land development and high storm runoff during rainy season, such is experienced in a country like Malaysia.

About 1.2 billion people still lack safe drinking water and more than 6 million children die from diarrhea in developing countries every year. In many parts of the world, river water that can be highly turbid is used for drinking purposes. World Health Organization (WHO) has set the guideline value for the residual turbidity in drinking water at 5 Nephelometric Turbidity Units (NTU) [1]. As identified by the United States Environmental Protection Agency (USEPA), turbidity is a measure of the cloudiness of water; it is used to indicate water quality and filtration effectiveness. Higher turbidity levels are often associated with higher levels of disease-causing micro organisms such as viruses, parasites and some bacteria. These organisms can cause symptoms such as nausea, cramps, diarrhea, and associated headaches [2].

Water-borne infectious disease caused by viruses, bacteria, protozoa and other micro organisms is associated with outbreaks of and background rates of disease in developed and developing countries worldwide [3].

Naturally occurring coagulants are usually presumed safe for human health. Earlier studies have found the Moringa oleifera seeds are non-toxic, and recommended its use as coagulant in water treatment in developing countries.

Moringa oleifera is the best natural coagulant discovered so far that can replace aluminum sulphate (alum), which is used widely for water treatment around the world.

The consumption of alum is very high in water treatment in Malaysia. The main concern in this study is low turbidity water which was difficult to be treated according to other researcher's presentations. The ideal plant for low turbidity water treatment in Malaysia is Wangsa Maju Water Treatment Plant (Puncak Niaga (M) Sdn Bhd) which was used for comparison in this study, the process flow sheet (Appendix A). The plant is working to treat the low turbid water with initial turbidity of 30 NTU or less. The process depends on using several additives to get residual turbidity of less that 5 NTU. The dose of alum usually used is between 13-18 mg/L, chlorine is added with an amount of 3.55 Kg/hour, and for post chlorination 5.5 Kg/hour, besides adding fluoride 0.5-0.6 mg/L, lime ($CaCO_3$) for pH adjustment, and Chamfloc 151 sludge dewatering with an amount of 3% w/w. Some chemicals are local and some are imported (Personal communications). Therefore, this study focused on the treatment of low turbidity water.

MATERIALS AND METHODS

The proposed method to isolate and purify the bio-active constituents is the cross flow filtration method, which produced the natural coagulant with very simple technique (oil extraction; salt extraction; and microfiltration through 0.45 µm). Process flow chart is shown in Figure 1.

Preparation of Moringa oleifera seeds

Good quality dry seeds of Moringa oleifera were selected from the pods that were collected from Serdang, Selangor Darul Ehsan, Malaysia. The pods collected were allowed to completely dry on the tree (the brown colour pods) because the green pods do not possess any coagulation activity [4].

The pods length ranged between (40-60) cm, and each pod contained around (20-30) seeds. The seeds coat and wings were removed manually, followed by the grinding of the seeds into a fine powder using a domestic blender (National, MX-896TM), then sieving the ground powder through 250 μm sieve (Figure 2).

Oil Extraction

Prior to extraction of bioactive constituents through ion exchange resin, the ground and sieved Moringa oleifera seed powder with size of < 250 μm was defatted with hexane by using electro thermal soxhlet apparatus (ROSS, UK). This was done by weighing of 10 gm of Moringa oleifera seed powder and setting it in the thimbles of the electro thermal soxhlet extraction chamber, adding 170 ml of hexane in the heating chamber, then evaporating of hexane through three cycles each for 30 min to ensure the extraction of oil from the seeds, until the hexane became colorless, drying of Moringa oleifera cake residue from the soxhlet thimbles and weighing the dry sample [5].

The Moringa oleifera cake residue stock after oil extraction was used in this study; the oil percentage was 35 % w/w. The sample of 1 Kg was defatted and kept to be used throughout the study period at room temperature a round 23±2 °C.

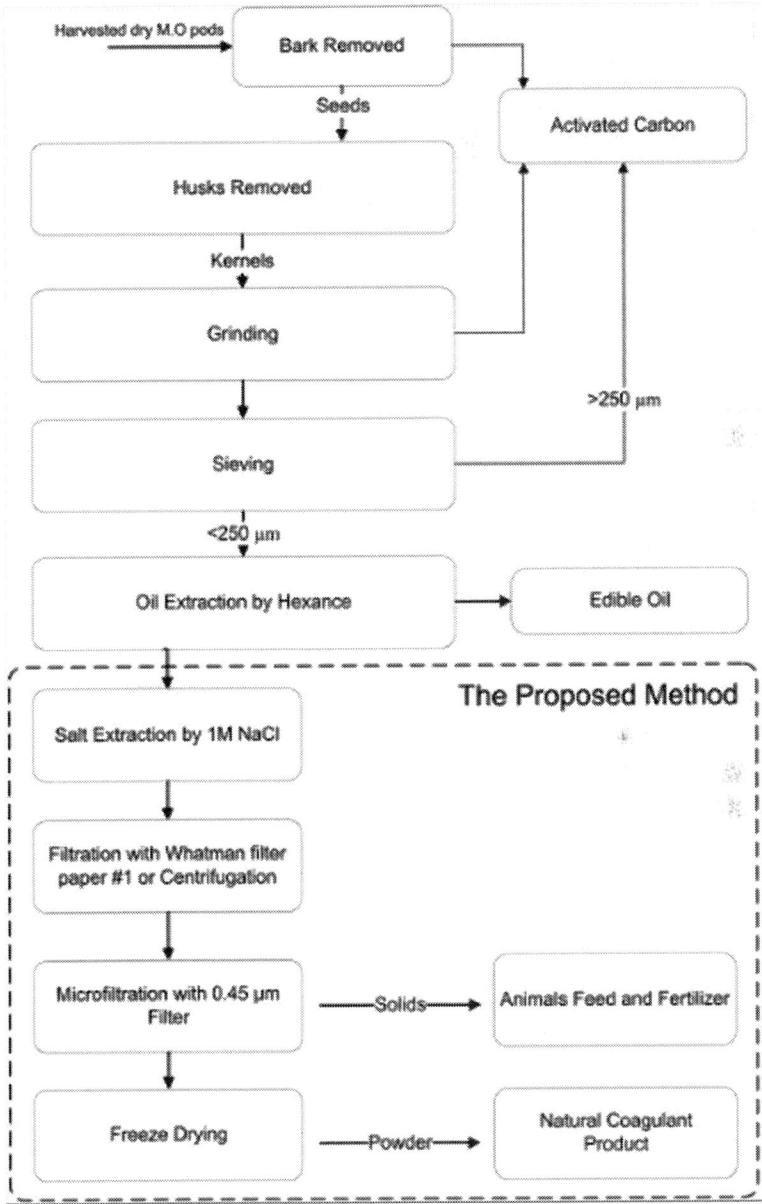

Figure 1: Proposed production process flow chart.

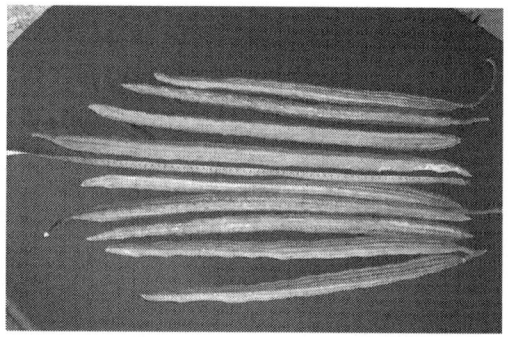

(a)

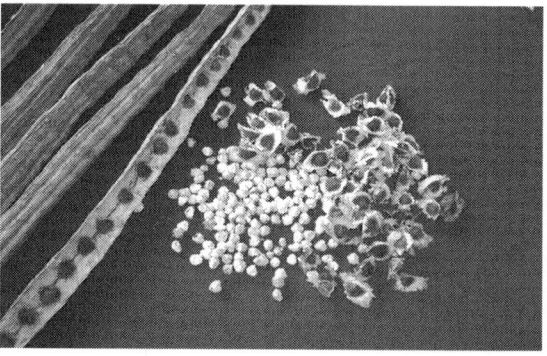

(b)

Figure 2: Moringa oleifera pods and seeds.

Salt Extraction

The extraction was done by adding a sloution of 1 Molar NaCl to the sample and mixed for 30 minutes using the magnetic stirrer (ERLA, ERLA Technologies (M) Sdn. Bhd. Malaysia). The extracted solution was centrifuged at 6000 rpm for a period of 10 minutes using the centrifuge (Eppendorf, 5804R, Germany). The supernatant was injected to ion exchange column.

Ion Exchange Process

The ion exchange technique was followed to isolate and purify the bioactive constituents for molecular weight measurements. The steps of the process were as follows:

Buffers Preparation

Two buffers were prepared for this stage of research as follows:

- Phosphate buffer stock (1 M): Phosphate buffer stock (1 M) with pH 7.5 was made by adding 174.16 gm of di-potassium hydrated phosphate (K_2HPO_4, GmBH Chemicals, Germany) into 900 ml of distilled water. The pH of the solution was 9.41 which were adjusted to 7.5 by adding 5 M HCl. The preparation of the buffer stock solution was finalized by adjusting the volume to 1000 ml by adding distilled water.

- Buffer A (0.1 M phosphate buffer): Preparation of 0.1 M phosphate buffer was by adding 100 ml of 1 M phosphate buffer stock into 900 ml of distilled water, 0.02 % NaN_3 was added to the solution as anti-bacterial agent.

- Buffer B (1.0 M NaCl in 0.1M phosphate buffer): This buffer was prepared by adding 58.44 gm of NaCl (GmBH Chemicals, Germany) to 900 ml of buffer A, and adjusting the volume to 1000 ml.

Preparation of the Column

The Glass Column (300×25 mm, with Bio-Rex Resin 70, Bio-Rad, USA) was used for ion exchange. The resin (Bio-Rex 70) was prepared by adding the resin powder to buffer A, and mixing the contents until the slurry was homogenous. Afterwards, the slurry was loaded to the glass column. The resin was then washed with 2 M NaCl to remove any impurities that may be present in the resin. It was then washed with buffer A to equilibrate the column to be ready for the loading of the sample.

Loading of the Sample to Ion Exchange Column

Equilibrate the column with 300 ml of 0.1 M phosphate buffer. Loading 30 ml of the extracted proteins sample as prepared in 2.3 above. Applying buffer A with an amount equal to twice of the column volume (300 ml) to wash all the unwanted negative charged proteins. Applying buffer B with 1 M NaCl, and buffer (A and B) and mixing them before sending to ion exchange. The flow rate was adjusted at 2 ml/min, and the fractions collected were 120 fractions with 5 ml volume each. Measuring absorbance of fractions using spectrophotometer (SECOMAM, Anthelie Advanced, France) at wave length 280 nm. The absorbance for the eluted fractions is plotted and 11 points were chosen for molecular weight measurements as shown in Figure 3.

Determination of Molecular Weight

High performance liquid chromatography (HPLC) equipment was used. The bio-active constituents purified by chromatography technique were injected to HPLC to determine the molecular weight. Eleven points were selected from Figure 3 to measure the molecular weight of different eluted fractions. HPLC (Waters 1525 with Binary HPLC Pump, Waters 2487 Dual Absorbance Detector, and Breeze GPC software, from Waters, 34 Maple Street Milford, MA 01757, USA) was used to measure the molecular weight of the bioactive constituents with Gel Filtration Column 300×7.8 mm (Bio-Sil SEC-125) for molecular weight measurement besides guard column 80×7.8 mm (Bio-Sil 125(was connected before the column to the HPLC to protect the column. The standard used was Bio-Rad Gel Filtration Standard with Thyroglobulin, IgG, Ovalbumin, Myoglobin, and Vitamin B12. The molecular weight is 670, 158, 44, 17, and 1.35 kDa, respectively. The buffer used was sodium phosphate (GmBH Chemicals, Germany). Filter paper (Waters GHP 0.45 μm) was used to filter the sample before injecting into HPLC.

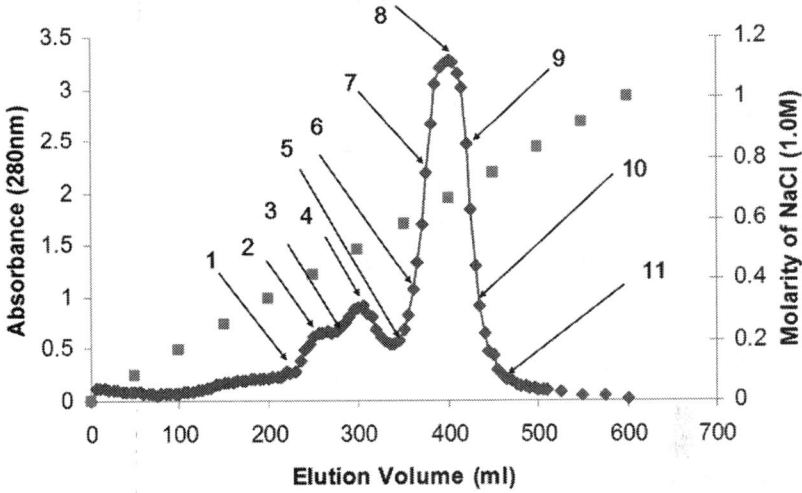

Figure 3: Eluted fraction from Ion exchange column.

The buffer used for HPLC was prepared as follows: 0.5 M NaH_2PO_4; 0.5 M Na_2HPO_4: 1.5 M NaCl; and 0.10 M NaN_3. The buffer used as a mobile phase for the HPLC is mixture from the above solutions as follows: 100 ml of 0.5 M NaH_2PO_4, 100 ml of 0.5 M Na_2HPO_4, 100 ml of 1.5 M NaCl, 100 ml of 0.10 M NaN_3. Subsequently by adding the above solution to one litre flask and completing the volume to 1000 ml by adding distilled water. The buffer prepared is contained the following concentration: 0.05 M NaH_2PO_4, 0.05 M Na_2HPO_4, 0.15 M NaCl, 0.01 M NaN_3 with pH = 6.8, and this Eluant is used as mobile phase in HPLC.

Cross Flow Filtration

The proposed method of microfiltration and ultrafiltration were performed for bioactive constituent's separation. The microfiltration followed by ultrafiltration was carried out as follows:

Microfiltration

The use of microfiltration was to reduce the organic load in the processed Moringa oleifera seeds. Cross flow filtration (QuixStand

Benchtop System, Sweden), with peristaltic pump (Watson-Marlow Bredel Pumps, Falmonth Cornwall TR 11 4RU, England) was used. The microfiltration cartridge (CFP-4-E-3MA) were used for sample filtration with pore size of 0.45 µm with a fiber ID 1mm, membrane area 110 cm², and nominal flow path length 30 cm. The structure is a polysulfone membrane which operates in a vertical orientation complete process fluid drainage. It is typical for concentration and purification of proteins, and suitable for scaling-up studies (GE Healthcare Bio-Science Corp. USA). The supernatant was then injected to ultrafiltration cartridge.

The new cartridge needs to be washed although it is shipped dry. It was washed with warm distilled water (55 °C) and rinsed for 5 minutes at a pressure of 0.3 bar (5 psig).

Ultrafiltration Process

The Xampler ultrafiltration cartridge (UFP-1-C-3M) was used for bioactive constituents separation with cutoff of 1000 Dalton, with fibre ID 0.5 mm, membrane area 140 cm², and nominal flow path length 30 cm, the type of membrane is polysulfone hollow fibre type (GE Healthcare Bio-Science Corp. USA).

Coagulation Activity Tests

The jar test was performed to monitor the coagulation activity of the processed Moringa oleifera seed in water treatment.

Synthetic Water (Kaoline Suspension) Preparation

A weight of 5 gm of Kaoline, laboratory grade (K7375, particle size 0.1-4 µm, Sigma-Aldrich), was dissolved with 500 ml of distilled water. Sodium bicarbonate solution with concentration of 100 mg/l was prepared by adding 100 mg of sodium bicarbonate (GmBH Chemicals, Germany) to 1000 ml of distilled water; Adding 500 ml of the sodium bicarbonate solution to the kaoline; Stirring the mixture at 200 rpm for 60 minutes to uniform the dispersion of clay particles. The suspension then was allowed to stand for 24 hours for complete hydration of the kaoline [4,6,7]. This is the stock kaoline solution for the coagulation activity evaluation throughout the study period and this stock was diluted a few times to get the turbidity needed for each particular test.

Jar Test

The jar test was performed for each stage of purification process to monitor the coagulation activity. The method applied was according to [5] with rapid mixing of 125 rpm for 4 minutes, followed by slow mixing of 40 rpm for 25 minutes and settling time of 1 hour. The jar test was carried out using (Stuart Flocculator SW6, Barloworld, UK) equipped with six paddles rotating in a set of six beakers. A turbidimeter 2100P (HACH company, USA) was used for all turbidity measurements.

RESULTS AND DISCUSSION

The molecular weight of the chosen eleven points for this study showed different low range molecualr weight (Table 1). All the points have low molecular weight which is ranged from 6500 to less than 1350 Dalton. The eluted fractions from the ion exchange column were used to monitor the coagulation activity and measuring the turbidity removal using the conventional jar test method. The results of turbidity removal percentage are tabulated in Table 2.

Table 1: Molecular weight for 11 points selected in Figure 3

Point #	Retention time (min)	Molecular weight Range (Dalton)
1	11.40	~4000
2	11.20. 11.50, 12.20	~5000, ~3500, < 1350
3	11.20, 11.40	~5000. ~ 4000
4	11.10, 11.40	~6000, ~4000
5	11.50, 13.20	~3500, < 1350
6	11.387, 12.167, 13.070	~4000, < 1350

7	11.10, 11.50, 12.20, 13.00	~6000, ~3500. < 1350
8	11.00, 11.40, 12.00	~6500, ~4000. 1500
9	11.50, 12.30, 13.05	~3500, < 1350
10	11.374, 12.162, 13.016	~4000. < 1350
11	11.383, 12.107, 13.110	~4000, < 1350

Table 2: Turbidity removal percentage for 11 selected points

Sample #	Optimum residual turbidity (N11J)	Turbidity removal %
1	10.62	64.60
2	2.23	92.13
3	2.95	90.16
4	1.78	94.08
5	2.03	93.24
6	1.38	95.39
7	5.21	82.63
8	4.71	84.29
9	1.65	94.49
10	0.64	97.12
11	1.56	94.80

The main concern in using Moringa oleifera for water treatment is the significant increase in organic load, as organic matters originating from the seeds can be released into the water during treatment [8,9]. The presence of organic matter in treated water can cause problems of colour, taste, and odour, and also facilitates the development of microorganisms upon storage [10]. Jahn [11] reported that water treated with crude Moringa oleifera extract should not be stored for more than 24 hours. The crude extract is therefore not generally suitable for large water supply systems where the hydraulic residence time is very high. Oil extraction is of a great advantage here, to reduce the organic load from the seeds, and to produce edible oil as a by-product. It is clearly possible to extract oil first and then use the aqueous extract as a coagulant. This dual exploitation is even advantageous for isolation and purifying the active agents in the coagulation with Moringa oleifera seed and also for the reduction of organic matter concentration in the treated water [4].

Two steps were followed to produce the natural coagulant by cross flow filtration. The first step was microfiltration, which is important for reducing the organic matters concentration. The main concern here is to filter all the oil from the seeds to reduce the organic content, because it can act as a precursor of trihalomethane formation during the disinfection process by chlorine which may be carcinogenic [8]. It is also may be postulated that the oil content in the seed will form an emulsion or film coating which may inhibit the contact with the surface of reaction and thus reduce floc formation. Extraction of the oil may therefore enhance the turbidity removal, resulting in better coagulation and flocculation [12].

The second step was ultrafiltration with 1000 Dalton cutoff membrane. In this study a simple scalable purification method was found. This is a straightforward production method of processed Moringa oleifera for water treatment.

The sample of Moringa oleifera seeds was processed as mentioned in Subsections 2.1, 2.2, and 2.3. The extract from the salt extraction process was filtered with Whatman filter paper #1 to remove all the solids from the extract. It was then applied for microfiltration with 0.45 μm microfiltration cartridge. Ndabigengesere & Narasiah [8] used 0.45 μm for microfiltration but the extraction method was with water, followed by many complicated steps to produce the coagulant, and the chemical oxygen demand increased by increasing the dose. No other research has been done on the microfiltration process. Permeate collected from microfiltration process was applied in the QuixStand Benchtop System apparatus to separate the retentate and permeate according to the molecular weight cutoff of 1000 Dalton.

The jar test was performed by using synthetic water and applying the three samples produced from: 1) microfiltration (0.45 μm), 2) the retentate of ultrafiltration (1000 Dalton), and 3) permeate of ultrafiltration (1000 Dalton), and the results of residual turbidity are shown in Figures 4, 5, and 6, respectively.

From the results obtained, 0.45µm microfiltration, showed a low residual turbidity of 1.32 NTU by adding 0.4 mg/L of processed Moringa oleifera to the synthetic water samples (Figure 4) with turbidity removal of 96.23 %. Also, there is no significant improvement by using ultrafiltration with 1000 Dalton after using microfiltration (Figure 5) because the retentate was 95 % of the total volume of the sample which is not economically feasible. By using a dose of 0.4 mg/L of the retentate, the residual turbidity was 2.29 NTU which is not much different from the result obtained from microfiltered sample through a 0.45 µm membrane. On the other hand, the use of permeate, gives residual turbidity of 2.38 NTU by using a double dose which is 0.8 mg/L (Figure 6) compared to the retentate dosage. This means that the bioactive compounds were passing to the permeate. Furthermore, for the ultrafiltration, it was observed that the bioactive constituents were covering all the molecular weights presented in the sample, because 5% of permeate of ultrafiltration gave good coagulation activity only by using the double dose (0.8 mg/L gave 93.20 removal efficiency and 0.4 mg/L of retentate gave 93.46 % removal efficiency). The result indicated that there was no improvement in the product quality using 1000 Dalton ultrafiltration membrane.

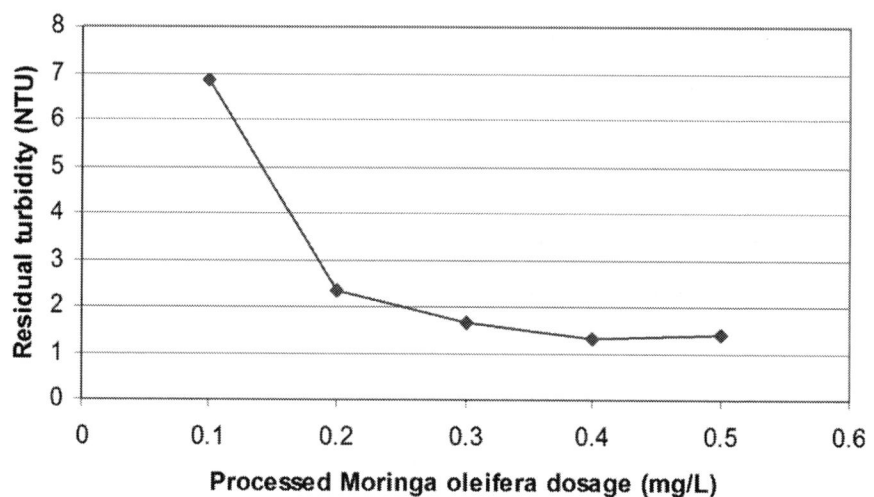

Figure 4: Residual Turbidity for microfiltration with 0.45 µm.

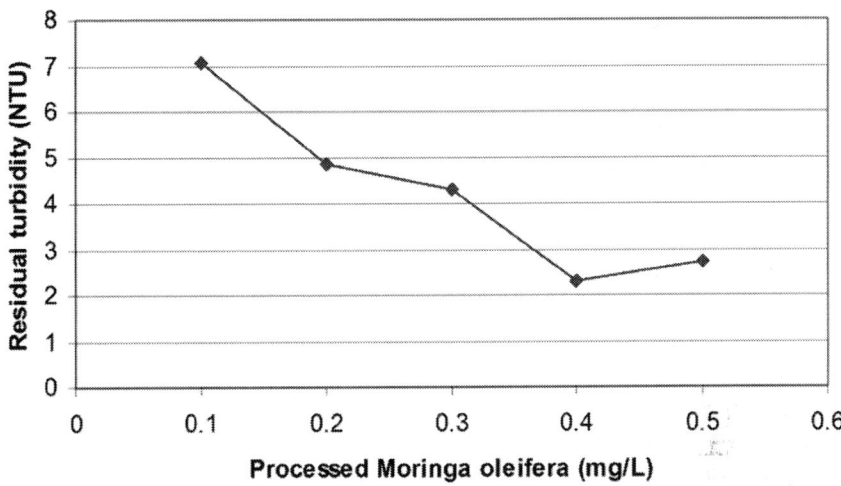

Figure 5: Residual Turbidity for microfiltration with 0.45 μm followed by 1000 Dalton (Retentate).

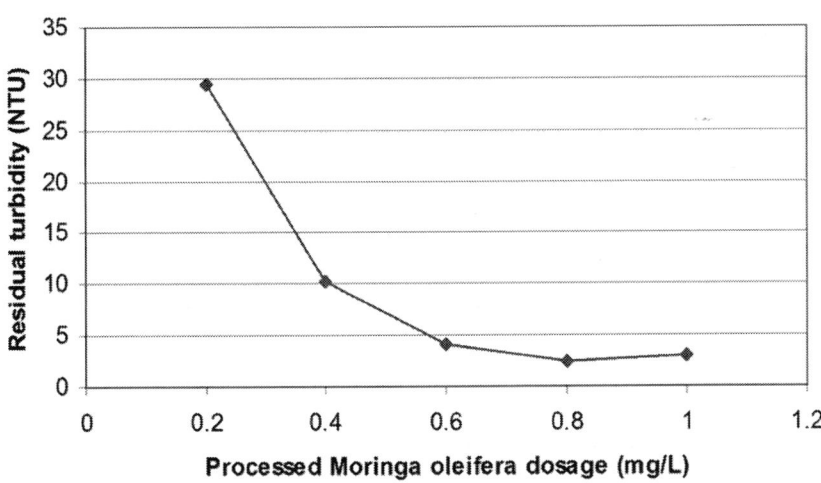

Figure 6: Residual Turbidity for microfiltration with 0.45 μm followed by 1000 Dalton (Permeate).

The ultrafiltration method involved the use of 1000 Dalton due to low molecular weight of less than 6500 Dalton (salt extraction method), in contrast with Ndabigengesere et al. [4] where the molecular weight was 13000 Dalton (water extraction method). The latter discovered out that the active components were retained on the 10000 Dalton membrane, while passed through the 30000 Dalton membrane.

CONCLUSIONS

The results showed that the cross flow filtration with microfiltration was sufficient to produce the natural coagulant with a more efficient and cost effective technique. It was observed that microfiltration with filter size of 0.45 µm is enough to produce a natural coagulant with good coagulation activity. It was clear in this study that to include ultrafiltration step in the production process was considered as a loss, besides the waste of time and labour. The ultrafiltration was not beneficial because permeate was very low, and separation was inadequate to improve the coagulation activity.

ACKNOWLEDGEMENTS

The authors are grateful to the Ministry of Science Technology and Innovation of Malaysia for funding the Project (IRPA 09-02-08-10002 EAR).

REFERENCES

1. G. L. McConnachie, G. K. Folkard, M. A. Mtawali, and J. P. Sutherland, "Field trials of appropriate hydraulic flocculation processes," Water Research, Vol. 33, No. 6, pp. 1425–1434, 1999.

2. http://www.who.int/water_sanitation_health/dwq/infectdis/en/index.html.Retrieved October 24, 2008.

3. http://www.epa.gov/safewater/contaminants/index.html. Retrieved October 24, 2008.

4. A. Ndabigengesere, K. S. Narasiah, and B. G. Talbot, "Active agents and mechanism of coagulant of turbid waters using Moringa oleifera," Water Research, Vol. 29, No. 2, pp. 703–710, 1995.

5.   S. A. Muyibi, S. A. Abbas, M. J. M. M. Noor, F. R. Ahmadon, "Enhanced coagulation efficiency of Moringa oleifera seeds through selective oil extraction," IIUM Engineering Journal, Vol. 4, No. 1, pp. 1–11, 2003.

6.   S. A. Muyibi and L. M. Evison, "Optimizing physical parameters affecting coagulation of turbid water with Moringa oleifera seeds," Water Research, Vol. 29, No. 12, pp. 2689–2695, 1995.

7.   T. Okuda, A. U. Baes, W. Nishijima, and M. Okada, "Improvement of extraction method of coagulation active components from Moringa oleifera seed," Water Research, Vol. 33, pp. 3373–3378, 1999.

8.   A. Ndabigengesere and K. S. Narasiah, "Quality of water treated by coagulation using Moringa oleifera seeds," Water Research, Vol. 32, No. 3, pp. 781–791, 1998.

9.   T. Okuda, A. U. Baes, W. Nishijima, and M. Okada, "Coagulation mechanism of salt solution extracted active component in Moringa oleifera seeds," Water Research, Vol. 35, No. 3, pp. 830–834, 2001.

10. M. Broin, C. Santaella, S. Cuine, K. Kokou, G. Peltier, and T. Joët, "Flocculant activity of a recombinant protein from Moringa oleifera Lam Seeds," Applied Microbiol Biotechnol, Vol. 60, pp. 114–119, 2002.

11. S. A. A. Jahn, "Using Moringa oleifera seeds as coagulant in developing countries," J. Am. Wat. Wks Ass., Vol. 6, pp. 43–50, 1988.

12. S. A. Muyibi, M. J. M. M. Noor, T. K. Leong, and L. H. Loon, "Effect of oil extraction from Moringa oleifera seeds on coagulation of turbid water," Environment Studies, Vol. 59, No. 2, pp. 243–254, 2002.

Chemical and Geological Control on Surface Water within the Shade River Watershed in Southeastern Ohio

Prosper Gbolo[1] and Dina L. López[2]

[1]Geology and Geological Engineering Department, University of North Dakota, Grand Forks, USA;
[2]Geological Sciences Department, Ohio University, Athens, USA

ABSTRACT

The under-sampled middle and western branches of Shade River Watershed (SRW) in SE Ohio were investigated as part of the Ohio University—US Environmental Protection Agency (EPA) STAR grant. This project was for monitoring the quality of watersheds in Ohio and classifying them according to their physical, chemical, and biological conditions. Water samples, as well as field parameters, were taken at twenty-two sites for chemical analyses. The ions analyzed included

Ca, Mg, Na, Fe, Mn, Al, NO_3, SO_4, HCO_3, and total PO_4, while the field parameters measured included pH, dissolved oxygen (DO), total dissolved solids (TDS), electrical conductivity (EC), and alkalinity. To assess the water quality within the SRW, the analyzed ions and field parameters were compared to the USEPA criteria for the survival of aquatic life. Analytical results showed that the watershed is dominated by $Ca-HCO_3$ waters with DO, Fe, Mn, and PO_4 being the main causes of impairment within the streams. The relatively elevated concentrations of manganese and less extent iron may be associated with the local geology and the acidic nature of the soils. The high alkalinity and calcium concentrations are due to the limestone geology. The elevated phosphate concentration may be due to anthropogenic sources, fertilizers, or contributions from phosphorus-rich bedrock that differs geochemically from other areas.

INTRODUCTION

Stream water and streambed sediment chemistry are continuously affected by the activities of man. Most of the water bodies in Ohio have been affected by natural and human factor that affects the physical, chemical, and biological quality of surface and groundwater [1]. Some of the natural processes include geology, soil, ecology, climate and physiography [1] and the human factors or activities include agriculture, urbanization, mining, and road construction, which have created instability in the ecosystems [2].

Agricultural activities have resulted in elevated concentrations of nutrients, sediment loads, and other nonpoint sources in streams [3, 4]. During storm water runoffs, most of the nutrients are transported in soils, and surface water. These nutrients are bio-accumulated in wetlands [5], sequester in soils [6-8] and incorporated in biomass [9] and streams or lakes. Excessive amount of the nutrients reduces the concentration of dissolved oxygen (DO) in surface water, leading to eutrophication [10- 14], and organic-rich sediments, which can directly or indirectly affect and alter the ecology of aquatic systems [15]. Low DO in streams can stress aquatic organisms, and this can lead to a high mortality rate or low diversity [16, 17].

The second factor affecting stream water quality in Ohio is urbanization. Currently, urbanization is second to agriculture as the main cause of stream impairment [18]. Due to population growth and urbanization, land surface are paved into impervious surfaces preventing infiltration of precipitation into the subsurface. This has created high erosion rates, large sediment loadings, and high flood frequencies. The transportation of sediments and high erosion rates have changed the morphology and hydrology of most streams [18, 19], thereby affecting the activities of benthic and aquatic organisms. The behaviors of aquatic invertebrates and macro invertebrates have been examined extensively in relation to pollution [20-23].

Coal mining is the third factor that affects water quality through environmental pollution and land degradation problems such as mine drainage and mine subsidence [24-26]. Mine discharge has resulted in high concentrations of acidity, iron, manganese, aluminum, and sulfate [27-29] in surface water bodies. The concentration of these elements can have detrimental effect on fish and macro invertebrate communities [30]. Studies on fish and macro invertebrate environments have revealed that acid mine drainage has affected fish and macro invertebrate communities [31], isolating populations in healthy portions of the streams.

This manuscript focuses on a study conducted to examine the surface water chemistry of the Shade River Watershed (SRW). The watershed is associated with high sedimentation rate, erosion and flooding [32, 33]. The objectives of this study are to: 1) determine the causes of impairment in the streams; 2) study the geological factors controlling stream water chemistry; and 3) understand the spatial distributions of the stream water chemistry, and mineral stability of the dominant ions within the SRW.

This study is part of the U.S. Environmental Protection Agency (EPA) Science to Achieve Results (STAR) project that focuses on land use, biology, and water quality within the Western Allegheny Plateau in Ohio. The purpose of the project is to classify streams according to their chemical, hydrological, geomorphologic, and biological gradient conditions. These conditions are very important for the sustainability of the aquatic environment.

Any variation in these conditions can have adverse effects on the water quality, physicochemical and biological components of the streams. Therefore, it is important to monitor the chemical properties of both the streambed sediments and the stream water, and compare them to the biological indicators [2].

MATERIAL AND METHODS

Study Area

This study was conducted within the SRW, a sub-watershed of the Ohio River within the Western Allegheny Plateau (WAP) in southeastern Ohio. SRW falls on the boundary between the southern part of Athens County and the northeastern part of the Meigs County (Figure 1). The Shade River drains an area of 570 square Kilometers and it flows into the Ohio River in the southeast. SRW is drained by the eastern, middle, and western branches of the Shade River [32, 33], but this study is restricted to the western and middle branches of the river. These two branches have not been extensively sampled prior to this study.

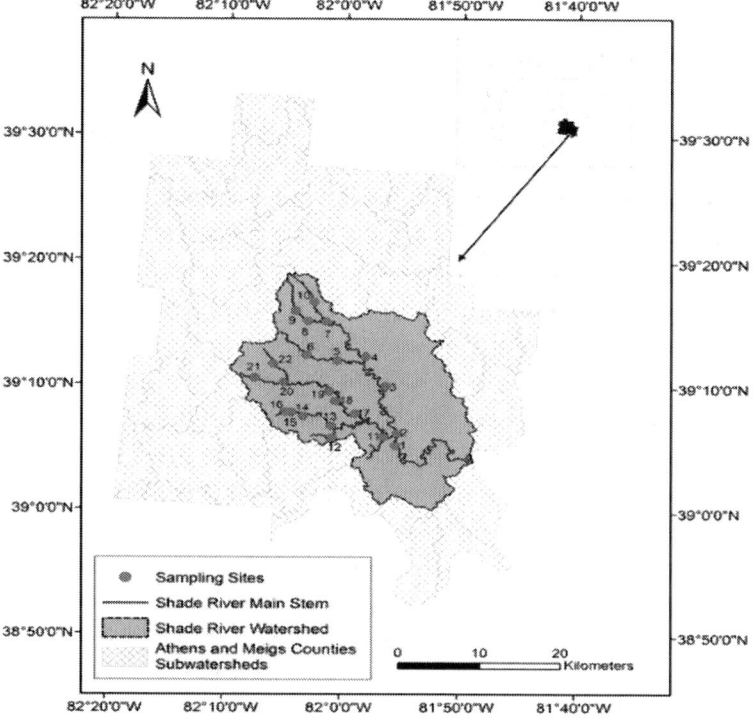

Figure 1: Map of the study area showing the sampled sites within Athens and Meigs Counties.

S According to Childress and Jones [33], part of the western branch of the Shade River was surface mined for coal during the mid-1940s to the early-1960s. This resulted in elevation of sediment loadings, and severe surface water and soil quality deterioration. Several attempts were made at that time to reclaim some of the mined and abandoned mined areas, but the efforts yielded no results. However, in 1978, new reclamation techniques were used to improve the sedimentation and water quality problems due to the passage of the Federal Surface Mining Control and Reclamation Act of 1977. Currently lime and other types of alkaline are used as buffers in curbing the water quality issues.

SRW is made up of rocks of the Conemaugh (western part of Meigs County), Dunkard, and Monongahela Formations (central part of Meigs County) of the Pennsylvanian [34]. These formations consist of sandstone, siltstone, limestone, shale and coal with some minor amount of conglomerate. The area is characterized by soils of the Gilpin-Upshur-Lowell-Guernsey association [35]. This soil association is moderately to well-drained soils that formed in colluvium and/or residuum from shale, siltstone, limestone, and some sandstone.

Sample Collection

Field sampling for this project was done in 2006 for grab stream water, as well as measured field parameters. Twenty-two referenced sites were selected and sampled based on a Geographic Information System (GIS) data generated according to USEPA criteria (Figure 1) [36].

The field parameters, which included temperature, dissolved oxygen (DO), electrical conductivity (EC), total dissolved solids (TDS), pH, acidity, and alkalinity, were measured at the time of water sample collection (Table 1). Temperature and pH were measured using the YSI Model 60 Handheld pH/temperature meter; EC and TDS were measured using CO150 model HACH conductivity meter; DO was measured using the YSI Model 55D Handheld DO meter; and alkalinity and acidity were measured using bromcresol green-methyl red and phenolphthalein indicators, and sodium hydroxide and sulfuric acid for acidity and alkalinity titrations, respectively.

Table 1: Physical and field parameters measured in the Shade River Watershed

Stream Name	Site	Lat.	Long.	T	pH	TDS	EC	DO
Sh. Rr.	1	39.08737	–81.92498	25.2	7.4	158	335	5
Middle Branch Sh. Rr.	2	39.10389	–81.92284	22.7	7.9	255	509	5
Middle Branch Sh. Rr.	3	39.16705	–81.94020	24.4	8.2	238	495	13
Middle Branch Sh. Rr.	4	39.20634	–81.96862	21.5	7.7	250	482	5
Pratts Fork	5	39.20031	–82.00905	22.7	7.9	225	448	4
Pratts Fork	6	39.20800	–82.05269	23.0	7.8	227	460	10
Middle Branch Sh. Rr.	7	39.25207	–82.02538	22.8	7.5	220	441	4
Middle Branch Sh. Rr.	8	39.25295	–82.05099	23.4	7.4	207	418	4
Middle Branch Sh. Rr.	9	39.26630	–82.06830	21.2	7.7	230	441	4
Long Run	10	39.27881	–82.04310	23.9	8.0	245	500	3
Walker Run	11	39.09914	–81.94148	22.3	7.5	180	359	5
Peach Fork	12	39.09749	–82.01529	23.2	7.6	163	326	3
Kingsbury Creek	13	39.11315	–82.01635	22.8	6.9	164	336	5
Kingsbury Creek	14	39.12622	–82.05617	23.3	7.6	170	345	5
White Oak Trib to Kg. Cr.	15	39.13135	–82.07428	22.5	7.6	158	311	5
Kingsbury Creek	16	39.13124	–82.07647	24.0	7.3	212	437	4
Trib. To West Br Sh. Rr.	17	39.12980	–81.98368	19.5	7.3	183	338	6
Trib. To West Br Sh. Rr.	18	39.14703	–82.01043	22.2	7.9	276	543	9
West Branch Sh. Rr.	19	39.15989	–82.02017	24.5	7.3	190	430	4
West Branch Sh. Rr.	20	39.17111	–82.08354	21.7	7.0	207	405	6
West Branch Sh. Rr.	21	39.17699	–82.12611	21.6	7.4	267	516	4
Trib. To West Br Sh. Rr.	22	39.19614	–82.09972	22.5	7.7	171	339	12

TDS and DO were measured in mg/L; pH (unitless); EC (μS/cm); and Temperature (T) in °C. Kg. Cr. represents Kingsburg Creek; and Sh. Rr. represents Shade River.

TDS and DO were measured in mg/L; pH (unitless); EC (μS/cm); and Temperature (T) in °C. Kg. Cr. represents Kingsburg Creek; and Sh. Rr. represents Shade River.

Filtered and unfiltered grab water samples were collected midstream for cations and anions analyses. The filtered samples were passed through a 0.45 μm millipore filter membrane using the vacuum filtration method and placed in a sterilized 250 milliliter (mL) high density polypropylene sample bottles for chemical analyses. The samples for cation analyses were preserved with diluted 2% nitric acid, while those for anion analyses were unpreserved. Samples for total phosphate analysis were preserved with 20% by volume sulfuric acid (H_2SO_4) and kept in 125 mL high density polypropylene sample bottles. The preservation of the samples was done to prevent the formation of metal complexes in the case of the cations and phosphate complexes in the case of phosphate. The samples were stored in a cooler filled with ice prior to standard analytical procedure. The cations were analyzed using Varian 720-ES Inductively Coupled Plasma Optical Emission Spectroscopy (ICPOES); expect Na, which was analyzed using the Shimadzu AA-6800 Atomic Absorption Spectrophotometer (AAS). The other cations analyzed included Na, Ca, Mg, Fe, Mn, and Al.

Sample Analysis

The anions were analyzed using HACH DR/2010 Portable Spectrophotometer [37]. The anions analyzed include Nitrate-N (NO_3-N), sulfate (SO_4), bicarbonate (HCO_3), and total phosphate (PO_4, Table 2). Nitrate-N concentration was determined using USEPA approved method 8192 called the Cadmium Reduction Method. Sulfate concentration was determined using USEPA approved method 8051 called the Turbid metric or SulfaVer4 method. Total phosphate concentration was determined using USEPA approved method 8180 called Hydrolyzable Digestion and the PhosVer3 or Ascorbic Acid Method. Bicarbonate concentrations were calculated from the measured alkalinity values obtained.

Statistical Analysis

Correlations between the chemical species are shown in Table 3. At the 95% confidence level (r = 0.05) considering 22 number of samples, the calculated t-critical and r-critical value for the test of significance of the correlation coefficients are 2.08 and 0.42 respectively. It can be observed in Table 3 that there is a positive and significant correlation between Mg and Ca (r = 0.74). Na correlates significantly with Ca (r = 0.60), Mg (r = 0.58), and Fe (r = −0.46). These correlations indicate a common rock source or similar chemical processes for the forma

Table 2: Concentration of some chemical species measured in the filtered water samples

Site	NO$_3$	SO4	T.PO$_4$	Alk.	Na	Ca	Mg	Fe	Mn	Al
1	0.11	9	0.43	165	94	103	23	0.13	0.35	0.31
2	0.10	11	0.22	173	93	121	29	0.09	0.14	0.36
3	0.06	14	0.32	162	18	90	21	0.03	0.14	0.19
4	0.12	10	0.18	186	19	88	20	6.30	0.13	0.19
5	0.12	13	0.27	141	97	112	28	0.13	0.15	0.49
6	0.07	9	0.33	163	19	84	18	0.08	0.15	0.20
7	0.13	4	0.21	160	99	118	30	0.14	0.24	0.26
8	0.31	4	0.44	167	111	110	28	0.23	0.35	0.25
9	0.11	14	0.22	186	83	102	22	0.09	0.16	0.18
10	0.08	23	0.24	143	74	109	29	0.07	0.12	0.19
11	0.23	11	0.29	201	115	98	26	0.23	0.23	0.20
12	0.10	7	0.21	159	96	103	24	0.19	0.22	0.21
13	0.10	19	0.19	93	87	104	22	0.10	0.19	0.35
14	0.28	11	0.19	160	86	107	23	0.15	0.26	0.17
15	0.14	29	0.43	117	61	88	17	0.12	0.20	0.19
16	0.06	14	0.55	85	70	126	23	0.11	0.33	0.23
17	0.05	11	0.31	153	84	108	27	0.08	0.11	0.16
18	0.10	11	0.25	145	69	116	32	0.10	0.11	0.18
19	0.15	27	0.20	44	94	100	25	0.10	0.43	0.15
20	0.07	21	0.31	143	100	131	27	0.05	0.47	0.18
21	0.03	16	0.17	180	105	127	29	0.13	0.19	0.23
22	0.04	4	0.30	149	85	100	20	0.37	0.14	0.27

T.PO$_4$ and Alk represent Total Phosphate and Alkalinity, respectively. All the elements are in mg/L except Alkalinity (Alk.), which is measured in mg Ca-CO$_3$/L.

T.PO$_4$ and Alk represent Total Phosphate and Alkalinity, respectively. All the elements are in mg/L except Alkalinity (Alk.), which is measured in mg CaCO$_3$/L.

Table 3: Correlation matrices for waters and field parameters within the Shade River Watershed. The highlighted values are statistically significant at 95% confidence level (|r| > 0.42; n =22 and = 0.05), as given by the test of significance of the correlation [38]

	Na	Ca	Mg	Fe	Mn	Al	pH	TDS	EC	DO	NO$_3$,	SO$_4$,	T PO$_4$,	Alk
Na	-1													
Ca	0.60	1												
Mg	0.58	0.74	1											
Fe	-0.46	-032	-026	1										
Mn	0.42	032	0.05	-0.19	1									
Al	0.28	022	0.17	-0.11	-0.13	1								
pH	-0.48	-0.34	-0.01	0.09	-0.64	0.04	1							
TDS	-029	030	0.46	022	-037	-0.002	0.50	1						
EC	-031	027	0.45	0.17	-0.27	0.005	0.51	0.98	1					
DO	*-035*	-036	-034	-0.05	-032	-0.13	039	0.11	0.11	1				
NO$_3$,	033	-0.14	0.08	0.03	033	-0.05	-0.09	-028	-025	-033	1			
SO$_4$	-0.07	-0.03	-0.13	-0.13	021	-022	-0.18	-0.05	0.01	-022	-0.15	1		
T PO$_4$	-0.08	0.00	-026	-023	033	0.02	-0.12	-027	-024	0.12	0.04	-0.02	1	
Alk.	-0.02	-0.08	0.11	0.24	-0.40	-0.004	038	028	0.15	0.11	0.13	-0.58	-0.19	11

tion of the aforementioned cations. The formation of the cations could be related to the weathering of carbonates or silicates. TDS correlates with EC (r = 0.98), as well as Mg (r = 0.45). This indicates that Mg is influential in the TDS concentration. pH correlates significantly with TDS (r = 0.50), and EC (r = 0.51) but correlates negatively with Na (r = −0.48), and Mn (r = −0.64) as shown in Table 3. The inverse correlation between pH and Mn indicates that as pH increases, the concentration of Mn reduces or vice-versa. Increase in pH in an oxidizing condition results in the precipitation of Mn [39] and manganese hydroxide [40] at alkaline pH. Sorption reduces the concentration of Mn in solution.

RESULTS AND DISCUSSION

Water Chemistry

The analyzed ions were used to determine the process controlling the water chemistry in the SRW based on the Gibbs [41] diagram. This diagram is based on three dominant processes: evaporation or crystallization, rock dominance or weathering, and precipitation. The graph of TDS versus $Na^+ / Na^+ + Ca^{2+}$ for the 22 data points indicates that the surface water within the watershed is located within the rock weathering portion of the boomerang-shaped graph (Figure 2).

Field Parameter and Water Chemistry Analysis

Stream water concentration within the SRW fell within the USEPA criteria for the protection of aquatic life except some few ions and field parameters that had higher values.

pH, TDS, EC, and alkalinity fell within the USEPA criteria except DO (Table 4). All of the sites had DO concentrations equal or greater than 4.0 mg/L, except Sites 12 and 14 that had values below the USEPA criterion of 4.0 mg/L [42] (Figure 3, Table 4). Iron concentration was within the USEPA criteria except site 4, which had a concentration of 6.30 mg/L (Table 4). The main causes of impairment within the SRW were total phosphate and manganese. The concentrations of these elements were beyond the USEPA criteria for the protection of aquatic life. The concentration of the total phosphate ranged from 0.2 to 0.5 mg/L (Figure 4, Table 4), which exceeded the USEPA criterion of 0.05 mg/L. Manganese on the other hand is associated with mining.

Manganese concentration ranged from 0.11 to 0.47 mg/L, which exceeded the USEPA criterion of 0.10 mg/L. Manganese concentration within sampled sites in the western branch was higher than those of the middle branch. The measured average concentration of manganese was 0.22 mg/L, which exceeded the USEPA criterion. Aluminum concentration ranged from 0.15 to 0.49 mg/L, with a mean value of 0.23 mg/L. Most of the areas associated with mining have high concentration of aluminum beyond the USEPA criteria of 0.2 mg/L. The high concentration of the aluminum could be due to acid mine drainage sources.

SRW had high concentration of the major cations (Ca, Mg, and Na), which could be associated with the weathering of carbonate or silicate minerals. The high concentrations of these elements depict the local geology of the All the elements are in mg/L except pH (unit less), EC (μS/cm), and Alkalinity (Alk.), which is measured in mg $CaCO_3$/L. *Most of the USEPA criteria values were obtained from [42].

Table 4: Descriptive statistics for the measured ions (filtered samples) and the field parameters with their respective criterion for aquatic life

Parameter	Minimum	Maximum	Mean	Standard Deviation	USEPA Criteria*
NO_3-N	0.03	0.31	0.10	0.12	10
SO_4	4	29	13	7.0	250
Total PO_4	0.17	0.55	0.28	0.10	0.10
Na	18	115	80	28	-
Ca	84	131	107	13	-
Mg	17	32	25	4	-
Al	0.15	0.49	0.23	0.08	-
Fe	0.03	6.30	0.41	1.31	1.0
Mn	0.11	0.47	0.22	0.11	0.05
pH	6.9	8.2	7.6	0.3	6.5 - 9.0
TDS	158	276	208.9	37.3	250
EC	311	543	418.8	72.2	-
DO	3	13	6	3	>4
Alkalinity	44	201	149	36	>20

All the elements are in mg/L except pH (unitless), EC (μS/cm), and Alkalinity (Alk.), which is measured in mg CaCO3/L. *Most of the USEPA criteria values were obtained from [42].

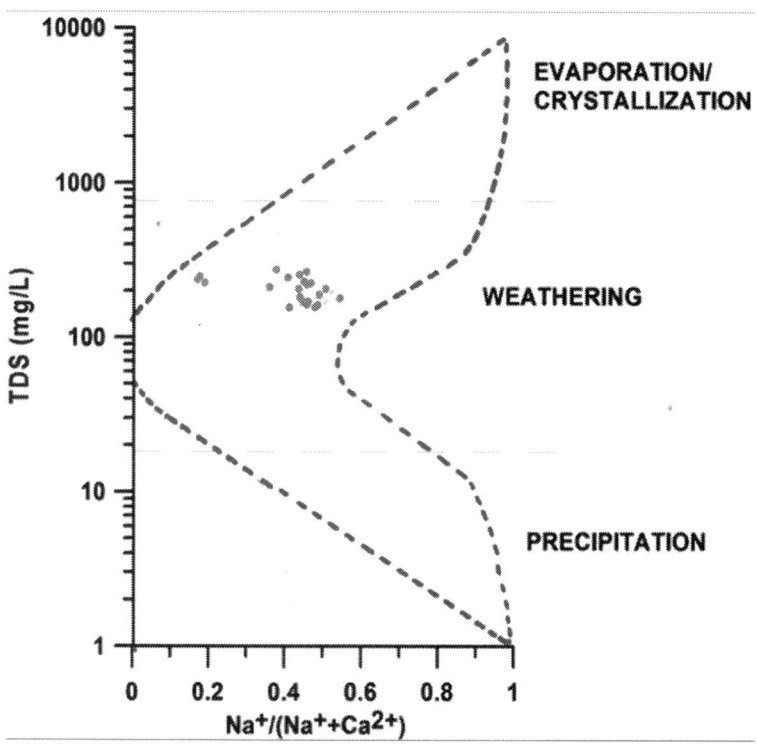

Figure 2: Dominant process controlling water chemistry within SRW [After 16].

area. The concentrations of the major cations were higher in the western branches as compared to the middle branches. The high concentrations of these cations could be attributed to the high rate of weathering, acidic nature of the soil, erosion of host rocks to the area, and/or from the remnant of the mining activities.

Composition of the Streamwater Samples

The composition of the surface water within the watershed was determined using ternary diagrams (Figures 5(a) and (b)). The diagrams indicate that the predominant cation in the waters is calcium (Ca^{2+}) while the predominant anion is bicarbonate (HCO_3).

Mineral Stability Phase

The SRW rocks contain carbonates consisting of calcite and dolomite. The weathering of carbonates, which depends on carbon dioxide and water, results in the formation of more calcium and bicarbonates ions. The equation for the weathering of carbonate is represented by:

$$CaCO_{3(s)} + H_2O_{(1)} + CO_{2(g)} \rightarrow Ca^{2+} + 2HCO_{3\ (aq)}^{-} \quad (1)$$

The equilibrium constant for calcite in Equation (1) is represented by:

$$K_{calcite} = \frac{a_{Ca^{2+}} \cdot a_{HCO_3^-}^2}{P_{CO_2}} \quad (2)$$

Where $^a Ca^{2+}$ and $^{a^2} HCO_3^-$ denote the activities of calcium and bicarbonate ions respectively (normally expressed in moles per liter of H_2O), while P_{CO_2} is the partial pressure of carbon dioxide ($10^{-3.5}$ atm at 25°C). A plot of the logarithms of $^a Ca^{2+}$ and $^{a^2} HCO_3^-$ shows that the surface water within the SRW is supersaturated with respect to calcite (Figure 6).

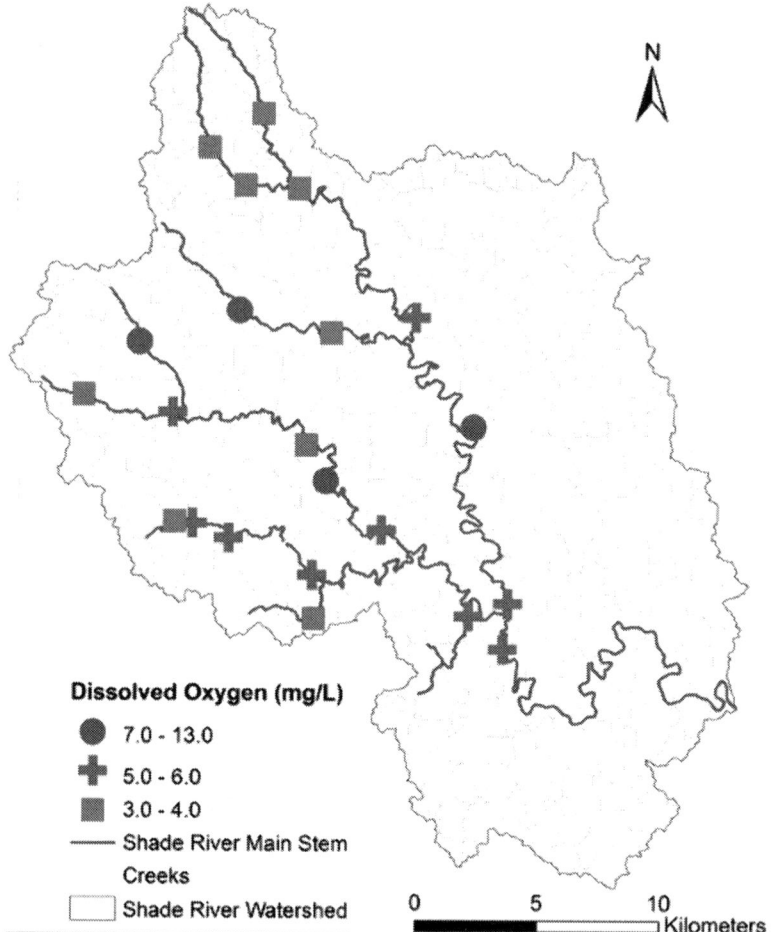

Figure 3: Spatial distribution of dissolved oxygen concentration in SRW.

Dolomite is formed when calcite dissolves in the presence of magnesium ions (Equation (3)).

$$2CaCO_{3(s)} + Mg^{2+}_{(aq)} \rightarrow CaMg(CO_3)_{2(s)} + Ca^{2+}_{(aq)} \quad (3)$$

To determine the stability field of the calcite within the area, the equilibrium reaction for the formation of dolomite was multiplied by the ratio of the square of the activities of the hydrogen ions to square of the activities of the hydrogen ions.

$$K_{Dolomite} = \frac{a_{Ca^{2+}}}{a_{Mg^{2+}}} \cdot \frac{a^2_{H^+}}{a^2_{H^+}}$$

(4)

Where $^aCa^{2+}$ and $^aMg^{2+}$ are the activities of calcium and magnesium ions respectively, while $K_{Dolomite}$ is the equilibrium constant for the formation of dolomite ($10^{0.2}$ at $25°C$). The stability fields for both dolomite and calcite for the waters in SRW was demonstrated by a line plot of $\log^a Ca^{2+}/a^2_{H^+}$ versus $\log^a Mg^{2+}/a^2_{H^+}$, as shown in Figure 7. The figure showed that the surface water in SRW fell within the stability field of calcite.

Mining Effect on Water Chemistry

The concentration of surface water collected near mined and reclaimed sites (Figure 8) show significant increase in the concentration of Al, Mn, and SO_4. These elements are associated with the problems of AMD. Iron, apart one site shows considerable reduction in the concentration due to the effect of lime used as a buffer in reducing the effect of AMD. The buffer effect of the lime and carbonates used can facilitate the precipitation of manganese hydroxide at alkaline pH under oxidizing condition [40].

CONCLUSIONS

Investigating the chemistry of surface water obtained from the Shade River Watershed has giving an insight to the impact of mining and other land use effect on water quality. Analyses of the various ions compared to the USEPA criteria for the protection of aquatic life has indicated that total phosphorus, manganese, and iron are the most impairing ions in the surface water. The water within the watershed is supersaturated with calcite and it is within the calcite's stability field.

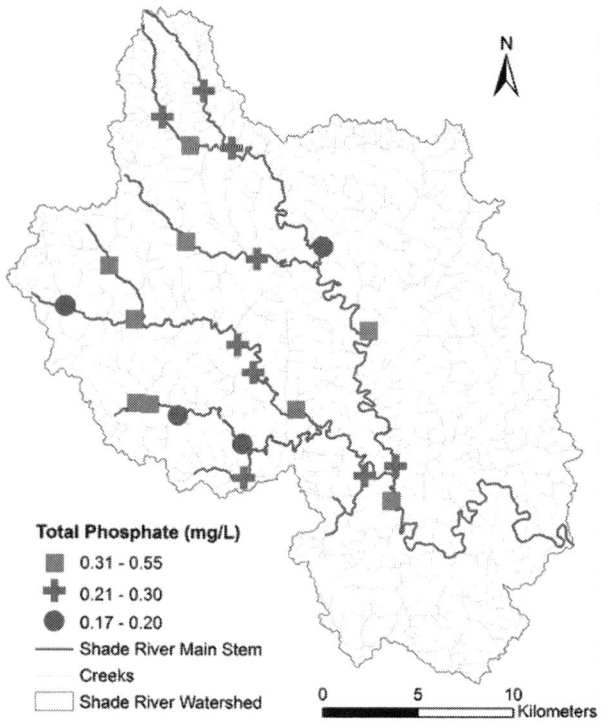

Figure 4: Spatial distribution of total phosphate concentration in SRW.

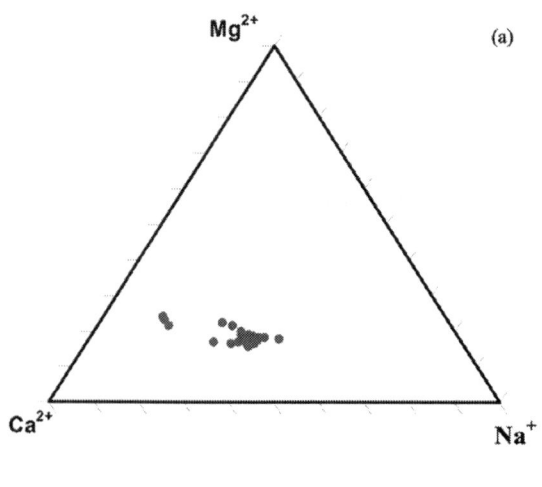

(a)

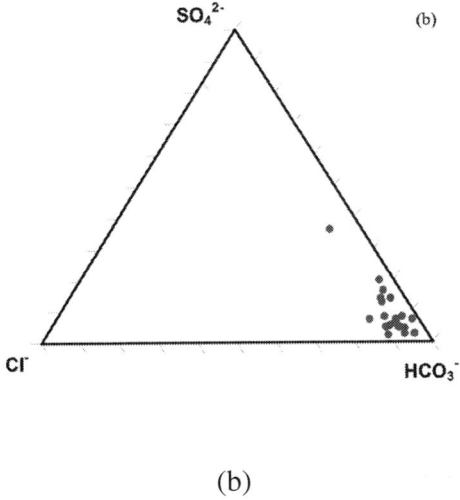

(b)

Figure 5: Ternary diagrams for the concentrations (milliequivalence in %) of the dominant ions in the SRW waters.

The high concentrations of some of the major ions depict the local geology and the anthropogenic influence on the watershed. Eutrophication is evident at some of the sites due to the high concentration of the total phosphorus from nasogical Sciences' Alumni Grant. The authors would like to acknowledge Mr. Kwarteng Amaning (formerly of Ohio University) for finishing Phase 1 of the project, and all the reviewers of this paper.

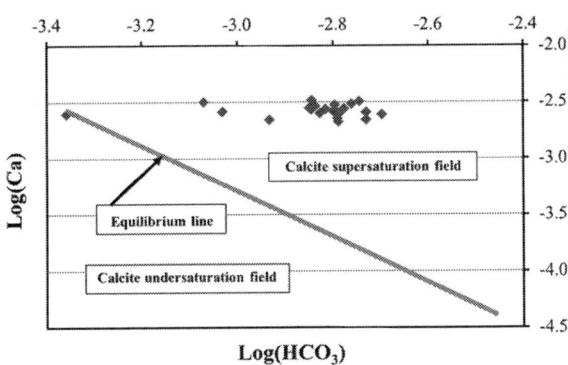

Figure 6: Phase diagram for the saturation field of calcite for surface water within the SRW.

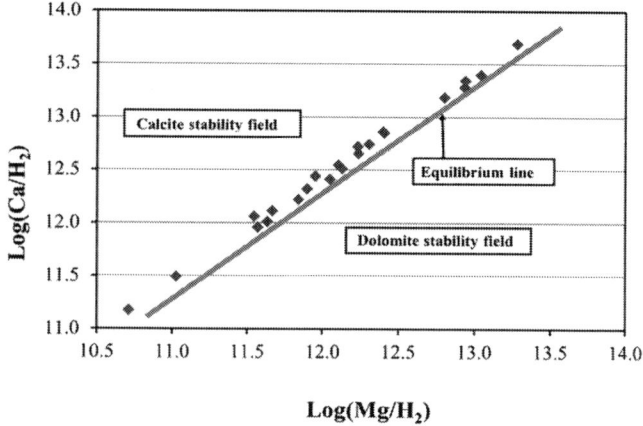

Figure 7: Stability field of calcite and dolomite for surface water within the SRW.

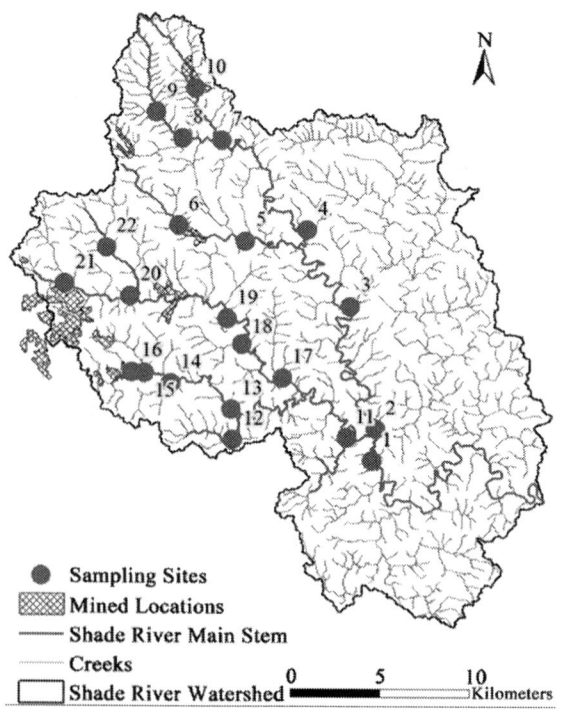

Figure 8: Map showing the mined and reclaimed sites in SRW.

REFERENCES

1.   L. M. Debrewer, G. L. Rowe, D. C. Reutter, R. C. Moore, J. A. Hambrook and N. T. Baker, "Environmental Setting and Effects on Water Quality in the Great and Little Miami River Basins, Ohio and Indiana," National WaterQuality Assessment Program, Water-Resources Investigations Report 99-4201, 2000.

2.   P. Gbolo, "Chemical and Geological Controls on the Composition of Waters and Sediments in Streams Located within the Western Allegheny Plateau: The Shade River Watershed," M.S. Thesis, Ohio University, Athens, 2008, p. 22.

3.   J. Hoorman, T. Hone, T. Sudman, T. Dirksen, J. Iles and K. R. Islam, "Agricultural Impacts on Lake and Stream Water Quality in Grand Lake St. Marys, Western Ohio," Water Air Soil Pollution, Vol. 193, No. 1-4, 2008, pp. 309- 322. doi:10.1007/s11270-008-9692-1

4.   E. D. Ongley, "Control of Water Pollution from Agriculture," FAO Irrigation and Drainage, Vol. 55, Food and Agriculture Organization of the United Nations, Rome, 1996.

5.   C. S. Hopkinson, "A Comparison of Ecosystem Dynamics in Freshwater Wetlands," Estuaries, Vol. 15, No. 4, 1992, pp. 549-562. doi:10.2307/1352397

6.   C. B. Craft, S. W. Broome and E. D. Seneca, "Nitrogen, Phosphorus and Organic Carbon Pools in Natural and Transplanted Marsh Soils," Estuaries, Vol. 11, No. 4, 1988, pp. 272-280. doi:10.2307/1352014

7.   C. Craft, "Freshwater Input Structures Soil Properties, Vertical Accretion, and Nutrient Accumulation of Georgia and US Tidal Marshes," Limnology and Oceanography, Vol. 52, No. 3, 2007, pp. 1220-1230. doi:10.4319/lo.2007.52.3.1220

8.   C. B. Craft and W. P. Casey, "Sediment and Nutrient Accumulation in Floodplain and Depressional Freshwater Wetlands of Georgia, USA," Wetlands, Vol. 20, No. 2, 2000, pp. 323-332. doi:10.1672/0277-5212(2000)020[0323:SANAIF]2.0.CO;2

9.   H. T. Stewart, P. Hopmans, D. W. Flinn and T. J. Hillman, "Nutrient Accumulation in Tress and Soil Following Irrigation with Municipal Effluent in Australia," Environmental Pollution, Vol. 63, No. 2, 1990, pp. 155-177. doi:10.1016/0269-7491(90)90065-K

10. J. E. Cloern, "Our Evolving Conceptual Model of the Coastal Eutrophication Problem," Marine Ecology Progress Series, Vol. 210, 2001, pp. 223-253. doi:10.3354/meps210223

11. S. W. Nixon, "Coastal Marine Eutrophication: A Definition, Social Causes, and Future Concerns," Ophelia, Vol. 41, 1995, pp. 199-219.

12. C. A. Scott, M. F. Walter, E. S. Brooks, J. Boll, M. B. Hes and M. D. Merrill, "Impacts of Historical Changes in Land Use and Dairy Herds on Water Quality in the Catskills Mountains," Journal of Environmental Quality, Vol. 27, No. 6, 1998, pp. 1410-1417. doi:10.2134/jeq1998.00472425002700060018x

13. A. N. Sharpley, T. Daniel, T. Sims, J. Lemunyon, R. Stevens and R. Parry, "Agricultural Phosphorus and Eutrophication," 2nd Edition, US Department of Agriculture, Agricultural Research Service, 2003, p. 44.

14. N. Wyngaard, L. Picone, C. Videla, E. Zamuner and N. Maceira, "Impact of Feedlot on Soil Phosphorus Concentration," Jounal of Environmental Protection, Vol. 2, No. 3, 2011, pp. 280-286. doi:10.4236/jep.2011.23031

15. D. Huggins and J. Anderson, "Dissolved Oxygen Fluctuation Regimes in Streams of the Western Corn Belt Plains Ecoregion," Kansas Biological Survey, 2005, p. 4.

16. D. J. A. Brown and K. Sadler, "Fish Survival in Acid Waters. In: Acid toxicity and aquatic animals. Society for Experimental Biology Seminar Series: 34, (R. Morris, et al., Eds.)," Cambridge University Press, Cambridge, 1989, pp. 31-44.doi:10.1017/CBO9780511983344.004

17. M. J. Paul and J. L. Meyer, "Streams in the Urban Landscape," Annual Review of Ecology, Evolution, and Systematics, Vol. 32, 2001, pp. 333-365.doi:10.1146/annurev.ecolsys.32.081501.114040

18. C. A. Burton, L. R. Brown and K. Belitz, "Assessing Water Source and Channel Type as Factors Affecting Benthic Macroinvertebrate and Periphyton Assemblages in the Highly Urbanized Santa Ana River Basin, California," American Fisheries Society Symposium, Vol. 47, 2005, pp. 239-262.

19. M. O'Driscoll, S. Clinton, A. Jefferson, A. Manda and S. McMillan, "Urbanization Effects on Watershed Hydrology and In-Stream

Processes in the Southern United States," Water, Vol. 2, No. 3, 2010, pp. 605-648. doi:10.3390/w2030605

20. J. Cortet, A. G. Vauflery, N. Poinsot-Balaguer, L. Gomot, C. Texier and D. Cluzeau, "The Use of Invertebrate Soil Fauna in Monitoring Pollution Effects," European Journal of Soil Biology, Vol. 35, No. 3, 1999, pp. 115-134. doi:10.1016/S1164-5563(00)00116-3

21. A. R. Gaufin and C. M. Tarzwell, "Aquatic Invertebrates as Indicators of Stream Pollution," Public Health Report, Vol. 67, No. 1, 1952, pp. 57-64. doi:10.2307/4587981

22. C. J. Goodnight, "The Use of Macroinvertebrates as Indicators of Stream Pollution," Transactions of the American Microscopical Society, Vol. 92, No. 1, 1973, pp. 1-13. doi:10.2307/3225166

23. R. W. Larimore, "Stream Drift as an Indication of Water Quality," Transitions of the American Fisheries Society, No. 110, 1974, pp. 627-637.

24. R. D. Hill and E. R. Bates, "Acid Mine Drainage and Subsidence: Effects of Increased Coal Utilization," Environmental Health Perspectives, Vol. 33, 1979, pp. 177-190. doi:10.1289/ehp.7933177

25. D. L. López and M. W. Stoertz, "Chemical and Physical Controls on Waters Discharged from Abandoned Underground Coal Mines," Geochemistry: Exploration, Environment, Analysis, Vol. 1, 2001, and pp. 51-60. doi:10.1144/geochem.1.1.51

26. E. Pigati and D. L. López, "Effect of Subsidence on Recharge at Abandoned Coal Mines Generating Acidic Discharge: The Majestic Mine, Athens County, Ohio," Mine Waters and the Environment, Vol. 18, No. 1, 1999, pp. 45-66. doi:10.1007/BF02687249

27. J. I. Sams and K. M. Beer, "Effects of Coal-Mine Drainage on Stream Water Quality in the Allegheny and Monongahela River Basins: Sulfate Transport and Trends," US Department of the Interior Water-Resources Investigations, Report 99-4208, 2000.

28. D. W. Schindler, "Effects of Acid Rain on Freshwater Ecosystems," Science, Vol. 239, No. 4836, 1988, pp. 149- 157. doi:10.1126/science.239.4836.149

29. J. G. Skousen and P. F. Ziemkiewicz, "Acid Mine Drainage Control and Treatment," 2nd Edition, West Virginia University

and the National Mine Land Reclamation Center, Morgantown, 1996, p. 362.

30. G. L. Lacroix, "Fish Community Structure in Relation to Acidity in Three Nova Scotia Rivers," Canadian Journal of Zoology, Vol. 65, No. 12, 1987, pp. 2908-2915.doi:10.1139/z87-441

31. M. W. Stoertz, H. Bourne, C. Knotts and M. W. White, "The Effects of Isolation and Acid Mine Drainage on Fish and Macroinvertebrate Communities of Monday Creek, Ohio," Mine Water and the Environment, Vol. 21, No. 2, 2002, pp. 60-72. doi:10.1007/s102300200021

32. C. J. O. Childress and R. L. Jones, "Sedimentation and Water Quality in the West Branch Shade River Basin, Ohio 1983 Water Year," US Geological Survey OpenFile Report 85-187, 1985.

33. C. J. O. Childress and R. L. Jones, "Sedimentation and Water Quality Data for the West Branch and East Branch Shade River Basin, Ohio 1984 Water Year," US Geological Survey Open-File Report 85-552, 1985.

34. R. A. Brant, "Geological Description and Effects of Strip Mining on Coal Overburden Material," The Ohio Journal of Science, Vol. 64, No. 2, 1964, pp. 68-69.

35. G. Gilmore and G. D. Bottrell, "Soil Survey of Meigs County, Ohio," United States Department of Agriculture, Natural Resources Conservation Service, 2000.

36. US Environmental Protection Agency, "Handbook for Sampling and Sample Preservation of Water and Wastewater," Environmental Monitoring and Support Laboratory, Office of Research and Development, 1982.

37. HACH, "Water Analysis Handbook," 4th Edition, 2005, pp. 31-200.

38. A. R. H. Swan and M. Sandilands, "Introduction to Geological Data Analysis," Blackwell Science, New York, 1995, p. 446.

39. D. M. DeNicola and M. G. Stapleton, "Impact of Acid Mine Drainage on Benthic Communities in Streams: The Relative Roles of Substratum vs Aqueous Effects," Environmental Pollution, Vol. 119, No. 3, 2002, pp. 303-315. doi:10.1016/S0269-7491(02)00106-9

40. A. W. Rose, B. Means and P. J. Shah, "Methods for Passive Removal of Manganese from Acid Mine Drainage," In: Proceedings of West Virginia Surface Mine Drainage Task Force Symposium, Morgantown, April 2003, p. 73.

41. R. J. Gibbs, "Mechanisms Controlling World Water Chemistry," Science, Vol. 170, No. 3692, 1970, pp. 1088-1090. doi:10.1126/science.170.3962.1088

42. US Environmental Protection Agency, "Quality Criteria for Water (Gold Book)," Regulations and Standard, Office of Water, Washington DC, 1986.

Chapter 6

The Environmental Dimension of Groundwater in Brazil: Conflicts between Mineral Water and Water Resource Management

Ana Lucia Desenzi Gesicki[1] and Francesco Sindico[2]

[1] National Department of Mineral Production (DNPM), Superintendence of São Paulo, São Paulo, Brazil

[2] Strathclyde Centre for Environmental Law and Governance (SCELG), University of Strathclyde, Glasgow, UK

ABSTRACT

There are three legal categories of groundwater in Brazil. Mineral water and potable table water are considered mineral resources, are part of the Federative Union's assets, and follow the legal regime applicable for the mining sector. "Normal" groundwater, on the other hand, falls under State's jurisdiction and is dealt within the Brazilian System of Water Resource Management, which promotes a decentralized and participatory management of water resources on surface or stored underground. This has led to conflicts of competences between federal (mining regulation) and state agencies (water resource management) because different concepts and styles of manage- ment are involved. This article argues for the urgent need to reopen the discussion aimed at a possible major revision of Brazilian mineral water legislation, in order to duly take into account the environmental dimension of groundwater as a public good for common use. An open-minded and transparent discussion involving the government and different sectors of society with com- peting interests in mineralized groundwater would therefore be highly recommended.

INTRODUCTION

In spite of Brazil's continental land area and abundance of fresh water, the country faces challenging tasks in managing water resources due to significant regional diversity and inequity of water distribution within its terri- tory. While the sparsely populated northern portion of the country encompasses the largest river basin in the world (the Amazon river basin), the Northeast suffers from periodic droughts and the Southeast has to tackle with the issues of water resources having an unbalanced supply-demand relationship and increasing pollution due to urbanization and industrial activities [1] [2] .

Regarding specifically Brazilian groundwater resources, these are subject to competing interests from various categories of use, mainly in the regions where the availability of surface fresh water of good quality is under threat from growing risk of pollution and scarcity due to critical climatic events. An overall estimate reveals that there must be about 400,000 wells across the country pumping groundwater for several purposes, at a likely total abstraction rate of 4000 m^3/s [3] . This estimate still gives a wide margin for safe water use considering that re- newable reserves of groundwater in Brazil have been estimated at about 42,000 m^3/s [4] .

The Brazilian Water Reform, carried out in the late 1990s, includes several principles and instruments inter- nationally accepted for dealing with water resources, such as integrated and decentralized management and mul- tiple uses of water. The water legislation, however, is silent on economic uses of mineralized groundwater. This is not surprising considering the aged legislation for mineral waters, namely Code of Mineral Waters (CMW), was enacted in 1945 and has never been updated. The CMW considers mineral water as a mineral substance with economic value and deals with it separately from water resource management. By doing so, mineral water falls under the same legal regime applicable for the mining sector. This has led mineral water and groundwater to be covered by different legal frameworks that embody distinct concepts and styles of management, which re- sults in conflicts of competences between federal (mining regulation) and state agencies (water resource man- agement).

Against this background, this article aims to critically assess the likely sources of conflicts between mineral water and water resource management, by providing an overview of both legal frameworks. This article is di- vided in six sections. The first three are devoted to provide an overview of the relevant aspects of the water, mining and environmental legislations in Brazil. The next two sections examine in detail the reasons why min- eral water legislation is in disagreement with the principles of the mining sector and of water resource manage- ment, including a reported case of conflict of interests between federal and local institutions. The last section fi- nally addresses key legal standpoints, which we consider important enough to justify a major revision of the current mineral water legislation.

WATER RESOURCES-PRINCIPLES AND POLICIES

Since the first decades of the 1900s, rivers have been extremely important in Brazil because of their huge poten- tial to generate energy. As a result, Brazil's energy matrix relies heavily on hydroelectricity. The Water Code of 1934 represents the early efforts to discipline the use of water resources in a country with an increasing demand for energy due to the growth of an urban-industrialized society [1] . During more than

60 years, the Water Code played an important role in governing the use of surface water resources for hydropower generation purposes, but several provisions of the law related to other matters (for example, groundwater use) have never been prop- erly regulated because the law did not take into account the complexity of different water uses. Although the Water Code is still in force, several provisions were implicitly revoked by the 1988 Federal Constitution. This has been the case for example, for water bodies previously ascribed as private property and municipal rivers.

The awareness for the need for environmental protection that emerged in the 1970s, combined with the further widespread understanding of the need to promote sustainable development, culminated in a change of paradigm from an ancient culture of disregard for available water to a new concept of water as a scarce resource [2] . The new environmental mind-set when dealing with water has led to a review of governmental strategies and actions for the implementation of an integrated management programme for water resources in Brazil [3] . The first step towards this achievement was the promulgation of the 1988 Federal Constitution itself, which assigned all waters to the public domain and mandated the implementation of the national system for water resource manage- ment. International legal instruments about water and the environment, such as the Dublin Declaration of 1992, reinforced the discussion about principles for water resource management and played an important role in the process of reviewing laws and policies at federal and state levels[1]. This led to the enactment of the National Water Act in 1997 (Law 9.433/1997), also known as "Water Law", which laid the basis for the National Policy for Water Resources and created the National System of Water Resource Management (SINGREH), launching new principles, objectives and administrative instruments for managing water resources in Brazil.

[1]Porto and Kelman cite that for the law-making process, several workshops took place during the first half of the 1990s, involving a broad set of participants, such as politicians, water professionals, NGOs and local communities in order to reinforce the need for decentralization of water management and the need for counting on democratic participation by society [1].

The National Policy for Water Resources is based on six principles that underpin the Brazilian system of wa- ter resource management.

First, water is considered to be a public good, following constitutional mandatory provisions. Second, it is recognized that water is a limited resource with economic value. Third, in the case of water shortage, the law establishes priority water uses for human and animal consumption. Fourth, the multiple uses of water must be considered within the management of water resources. Fifth, it establishes the river basin as the basic territorial unit for the implementation of the National Policy for Water Resources and the National System of Water Resource Management. Sixth, the law promotes the decentralized and democratic management of water resources, counting on the participation of governmental institutions, stakeholders and communities involved in the river basins.

An important advance in Water Reform was the introduction of a permit system, at a national level, to grant rights for any use of water resources stored underground. As groundwater is under the exclusive jurisdiction of the States[2], groundwater users must follow the rules established by the water permit system defined at State lev- el[3]. For granting water rights, some relevant issues are observed by the State bodies, such as the water availabil- ity of the aquifer in the river basin, its expected recharge rate and the stipulation of a maximum yield for groundwater abstraction that is considered "sustainable", i.e. the volume of water in a rate of abstraction that guarantees the availability of groundwater for other uses [5] . The authorization for water use issued by the State institution is time limited, a time limit that varies from State to State[4].

[2]Article 26 of the 1988 Federal Constitution.

[3]Water permit systems vary in national range because States are allowed to promulgate their own legislation of water resource management, following the guidelines established in the National Policy for Water Resources. For this reason, there are States with long-term and well established water permit systems (e.g. São Paulo and Minas Gerais States) and others which have just promulgated its own water legislation and still face difficulties with the enforcement of the law.

The maximum period of validity of water rights is 35 years.

The dual jurisdiction over water bodies has been identified as a challenge for the integrated management of water resources. That is the case of surface waters, which can be under the jurisdiction of the

Union or the States depending on their geographic localization[5]. It is not different for groundwater. Even though the Constitution defines the jurisdiction of groundwater to the States, there are not yet legal provisions to integrate mineralized groundwater management to the National Policy for Water Resources and so providing a desirable integration between federal and State agencies. For a country of great regional diversity, the governance of water resources depends primarily on the cooperation between the different levels of the government, which is not simple to achieve [3] considering that Union, States and Municipalities all have administrative autonomy to define rules and standards to deal with protection of the environment and other issues of their own competences.

[5]Surface fresh water bodies (rivers and lakes) are under the States' domain except if they serve as boundaries with other countries or between States, which are considered of federal jurisdiction.

PROTECTION OF THE ENVIRONMENT

At the time when the legislation for mineral water was drafted, Brazil had great potential for development and was starting to face a significant increase in the rate of industrialization and also undergoing a population explo- sion, mainly in the southern urban centres. Paradoxically, this accelerated economic growth was founded on in- creasing regional economic discrepancies, and occurred at the expense of individual freedoms and furthermore imposed environmental costs because of the uncontrolled use of the country's natural resources [6] -[8] .

The model of development at all costs started to be questioned in the 1970s, mainly as a result both of a movement of civil society organizations and of several serious environmental damages in Brazil[6]. Environmen- tal regulation at federal level in Brazil started with the promulgation of Law 6.938/1981, which established the National Environmental Policy, and set up the National System for the Environment (SISNAMA) multi-institu- tional framework. Within SISNAMA there are institutions and entities at Federal, State and Municipal levels that grant environmental licenses, carry out inspections and enact supplementary rules and policies within the ambit of their own competences in matters related to environmental protection and pollution control.

[6]One of the most serious events of environmental damage occurred in Cubatão, an industrial city in São Paulo State, where uncontrolled pollution caused by 24 petrochemical and steel companies resulted in mountain landslides, contamination of soil, rivers and air, as well as grave health problems to local population. The companies were involved in a lawsuit filed under the Public Civil Action conducted by the State Public Ministry from 1985 and obliged to fully remediate the environment [9].

Environmental regulation of economic activities in Brazil stems on two internationally widespread principles: the precautionary and the polluter-pays principles. The precautionary principle requires a proactive approach towards the protection of the environment, using two main instruments of the National Environmental Policy — environmental impact assessment (EIA) and licensing of economic activities that cause potential or effective hazards to the environment[7]. The environmental viability of a project is analyzed by SISNAMA bodies at Fed- eral, State or Municipal level, depending on the type of activity and the scale and localization of the project. The polluter-pays principle is expressed in the penalties, sanctions and requirements for reparations to which offend- ers are subject if they cause harm to the environment. Note that mining is an activity that admittedly implies damages to the environment; hence, the miner has the constitutional obligation to restore what has been de- graded.

[7]The construction, installation, expansion and operation of any potentially hazardous activity are subject to a three-step licensing process (preliminary, installation and operation licences). The preliminary, installation and operation licences of a project are time-limited, not exceeding five years, six years and ten years, respectively. The environmental licensing process eventually involves the realisation of public meetings for divulgation of the project to the affected community.

The environmental legislation, combined with the environmental principles of the Federal Constitution and of the Brazilian Civil Code[8], establishes water as a public good for common use, describing it generally as an en- vironmental resource, together with other elements of ecosystems. Environmental resources are for general use of the whole community, being essential to acquire and maintain a healthy quality of life and to reach the ideal ecological equilibrium. On the other hand, the private use of water resources through the permit system of the National Policy for Water Resources does not denote possession of the water itself [9] , considering the premise of inalienability of public goods for common use [10] . Rather it implies the right to use water for economic or individual purposes, observing the restrictions imposed by law and by the collective interests. Integrating the public domain with other environmental resources, water in general cannot be privately or individually owned. Rather the opposite since the public bodies with legal competence for managing public goods are bound by law to guard them (include water) for the overall benefit of the community [10] .

[8]Article 99 of Brazilian Civil Code (Law 10.046 of 10/01/2002).

Against this background, groundwater acquires a strong environmental dimension. As an environmental re- source, it constitutes as an important reserve of drinking water for future generations and must subject to protec- tive and conservative policies for keeping an ecologically balanced environment. The economic use of ground- water resources promotes either private or collective interests, supporting economic activities or supplying the population with essential public services. Nevertheless, the economic dimension of groundwater must be com- patible with the intergenerational obligation of guaranteeing groundwater reserves in adequate quality and quan- tity for the future.

MINING AND MINERAL WATERS-REGULATORY FRAMEWORK

The Code of Mineral Waters (CMW) was enacted while the Constitution of 1937 and the former Code of Mines of 1940 were in force. The Constitution of 1937 reiterated most of the principles for mineral resources exploitation introduced in 1934. According to it, the property of mines and other subsoil wealth was considered distinct from the ownership of the land, and their economic exploitation relied upon federal grants. These main principles were later reinforced by the Federal Constitution of 1988, currently in force, which extends the Federative Union's domain over mineral resources even for those undiscovered, wherever they are settled (on surface or in the subsoil), and also defines that mining activity must be carried out in the national interest, following policies and rules for protection of the environment and for the recovery of degraded areas by the mining company.

The CMW was promulgated in 1945 after the work of a Hydrology Commission nominated by the Ministry of Agriculture in 1940. The law clarifies that the economic exploitation of mineral waters is governed by the legal regime for mineral resources based on the provisions of the Code of Mines of 1940, subsequently updated by the Code of Mining (CM) of 1967.

Since 1967, the CM has governed all aspects of the mining sector in Brazil[9], from exploration of mineral re- sources to production and commerce of mineral commodities. Mineral rights are established and extinguished by successive administrative proceedings, mostly of a binding nature [11] , issued by both the Ministry of Mines and Energy (MME) and the National Department of Mineral Production (DNPM).

[9]By the time of drafting this article, 1967 Code of Mining was still in force. Nevertheless, the bill number 5.807/2013 has been in progress in the National Congress since 18/06/2013, which will update the cited law.

Authorization and concession are legal regimes defined by the CM that represent the two main phases of the mining activity, respectively the exploration and exploitation stages. The authorization regime comprises a time-limited title (2 to 3 years, renewable) that allows either individuals or companies to conduct, at their own expense, all prospecting works needed for the definition of an ore deposit, evaluation of mineral reserves and grades, as well as the determination of their economic feasibility. The concession title is issued by the MME, has unlimited duration and allows private companies to carry out mining operation projects that have been pre- viously submitted to an environmental licensing process.

The Constitution empowers the Federative Union to dispose of mineral wealth, be it on the surface or under- ground, without any legal restriction. The first competent applicant to request a survey grant for a particular available area can obtain exploration priority rights. The right of priority, also known as the "first come, first served" rule, is one of the mining activity principles that guarantees democratic access to mineral resources, as the Union must grant mineral rights to whoever first legitimately applies for it.

The mining sector has a peculiar dynamic as economic activity. Despite the public nature of mineral resources, mining is not a public service. All investments (normally of a long term nature) and usually high risks are at the expense of private parties. Once a mining title has been issued, the miner begins to have a right/duty to exploit mineral deposit until its depletion. The property of the mined product is guaranteed to him, as is the right to use, enjoy, sell or rent the mine and to use the mineral rights as collateral. Thus, the exploitation concession is the ad- ministrative act that allows the shift of the ownership of a public asset (in situ ore deposit) from the Union to private hands (mined ore deposit), incorporating it into their assets [12] [13] .

Mining is constrained by the immovable nature of the ore deposits. Mineral resources can be found anywhere, outcropping the surface or deep inside the soil. Considering the random and unpredictable natural occurrence of mineral resources, mining is considered to be of public utility or of social interest and direct exploitation in some protected areas can be allowed only under specific environmental law provisions[10].

[10]Resolution 369/2006 of the National Environmental Council (CONAMA) reinforces the public utility and the social interest of activities of exploration and extraction of mineral substances, by permitting eventual intervention or removal of wood cover in Permanent Protected Areas defined in the Brazilian Forest Code.

The Code of Mineral Waters (CMW)

Mineral water is legally defined as groundwater with a distinct chemical composition or physical-chemical properties which confer a medicinal action. The CMW recognizes two categories of groundwater: mineral wa- ters and potable table water. Both categories of groundwater are considered mineral substances and are governed by the mining legal framework. The CMW regulates two only possible end uses, for bottling process and for spas purposes.

Mineral waters are defined as having a medicinal action depending on their composition. They are characte- rized by a wide range and defined amounts of some chemical compounds (radium, sodium and calcium bicar- bonate, sulphate, sulphur, nitrate, sodium chloride and iron), by the presence of dissolved gases (carbon dioxide, hydrogen sulphide, and short half-life radioactive gases) and by the contents of noteworthy trace elements (iodine, arsenic, lithium, etc.). Together with chemical composition, some properties of the water at the source, such as temperature and temporary radioactivity, are also used to classify a mineral water source.

Potable table water, on the other hand, has nothing special at all, i.e. comprises groundwater which does not reach the chemical or physical parameters defined by the CMW, but only potability. Potable table water theoreti- cally is not associated with any sort of presumable beneficial health effect, but is suitable for bottling purposes.

According to the CMW, water with particular chemical and physical parameters is defined as having thera- peutic benefits, even though no therapeutic effects may have been shown empirically by medical experts. The medicinal action is the main legal criterion that defines a mineral water, but the procedure of testing for any fa- vorable effect on health is not a legal requirement[11], because the law sets down the chemical and physical para- meters for mineral waters (Art. 35 and Art. 36 of CMW).

[11]Groundwater classified as "oligomineral", "nitrated" and "chlorided" are kinds of mineral water that exceptionally must have its beneficial effect on health attested through empirical measurements carried out by specialized doctors, which results depending on approval of a board of experts in Crenotherapy. CMW, Art. 35, I together with Art. 1, §2 and §3.

The CMW was enacted at a time when several spas and hydromineral resorts were designated as important healing centres in Brazil. Mineral waters were normally known as "mineral-medicinal" or "thermal-medicinal" waters, because Crenotherapy was experiencing great popularity as a traditional practice in medicine [5] [14] . After the Second World War, however, the discovery of penicillin and the spread of the pharmaceutical industry have been identified as the main reason for the decline of those medicinal practices, not only in Brazil but also worldwide [15] . Bottled mineral waters have their origin in the spas and resorts, retaining a connexion to their sources and the brands that carried the prestige of some known effects on health, but slowly they became more and more separated and diverged from the resorts [15] , splitting up completely and developing as independent companies in a fast growing sector. The production of bottled mineral water before

1970s were lower than 100 million litres/year, increasing to around 800 million litres/year in 1990 and then boosting to more than 3000 million litres/year from 2000 forward [16] . By the end of the 1950s, bottled mineral water was no longer sold in pharmacies but had begun to be marketed in supermarkets, restaurants and bars [15] , and was seen not anymore as a medicine, but also as a domestic consumer good. The mineral water sector changed its focus from healing purposes, leaving the medical scope and approaching the bottled beverage industry, to dealing with the primary purpose of the water: thirst quenching.

Considering the extinction of the Crenotherapy specialty or subspecialty in medicine, the confirmation of some therapeutic use of a mineral water has not been made since the 1960s. In that way, the classification of mineral water has been based only on the chemical or physical properties defined by law. On the other hand, the legally defined parameters used to classify a mineral water cover such a wide range (for example the tempera- ture in the source) that practically any grade of mineralization or even waters containing low amounts of trace elements (for example fluoride) are enough to categorize them as mineral water. In practice, today, any naturally potable groundwater is likely to be classified either as mineral water or as potable table water [15] [17] if suita- ble for bottling or for spa purposes.

MINERAL WATERS VERSUS MINERALS-A COMPARISON

Mineral water seems an alien mineral substance within the mining sector and remarkable differences between mineral water and other minerals can be pointed out as follows: 1) mineral water is neither a mineral nor a fossil substance; 2) mineral water is a renewable resource, in theory; 3) mineral water is produced directly for human consumption, and because of that, the sector must follow standards and rules promoted by health, sanitation and environmental institutions; 4) inspection procedures for mineral water are more similar to the ones taking place in the sanitation and health sectors than in the mining one; 5) exploitation of mineral water implies a minimal alteration of the surface and minimized visual impacts at the point of extraction (punctual extraction); 6) mineral water exploitation is conservative because it involves a delimitation of a protection area around the source with restriction to any underground work or intervention; and 7) mineral water extraction and processing cause desirable minimal change in the original water composition.

The first basic difference between minerals and mineral waters refers to their nature. Water is not considered a mineral because it is liquid and does not show a crystalline structure [18] . Mineral water is not the only excep- tion within other mineral substances though. For example, mercury occurs as a liquid and opal is amorphous, so both are considered mineral-like substances, named mineraloid, but are not formally accepted as minerals. Some chemical compounds induced by biological processes, such as marine phosphorites, are not truly minerals either. In fact, being a mineral is not a sine qua non for being regarded as a mineral resource.

Another key difference regards renewability. Mineral water is a drinkable groundwater which, theoretically, is a renewable resource connected directly or indirectly to surface waters, taking part in the hydrological cycle. According to Rebouças, 97% of all groundwater is meteoric in origin and was recharged by infiltration of sur- face water into the ground at some moment in geological time [19] . The concept of renewability for groundwa- ter, however, is relative to time and can be divided into renewable and non-renewable. Renewable aquifers re- ceive natural recharge and have the ability to replenish their reserves depending on the level of connectivity with surface waters, in a period of time comparable to a human lifespan. On the other hand, aquifers that need a very long period (hundreds to thousands of years) to replenish their stocks are known as non-renewable or fossil [20] . Non-renewable groundwater is normally found in unconfined aquifers where contemporary recharge amounts are small or infrequent due to severe climatic conditions, as well as in confined sections of some aquifers where active recharge is negligible [20] [21] . "Groundwater mining" is the convenient term often used for groundwater developments that involve the abstraction from fossil or non-renewable water reservoirs. In other words, ground- water mining may be defined as the exploitation of groundwater at a rate that is much greater than the rate of replenishment of the aquifer storage [22] [23] .

Nevertheless, the principles that underpin the mining activity and water resource management are quite dif- ferent. One of the problems of mineral water management is the premise that mineral resources can be exploited until ores deposits are exhausted, because this principle implies some environmental concerns related to the pro- tection and conservation of groundwater as an environmental resource, which are not properly covered by the CMW or the CM. Because the CMW and the CM antedate the Brazilian environmental legislation, most of the policies and rules concerning environmental protection affect the mining sector directly without disrupting the mining legal framework. The environmental legislation introduced another step for the mining activity, by sub- mitting mining developments to environmental licensing process regarding its installation and operation. Consi- dering that mining is legally recognized as a potentially hazardous activity to the environment, mineral exploita- tion (including mineral water) is not authorized unless previous environmental impact assessment, or another similar instrument, has been carried out. The theoretical depletion of a mineral water "deposit" can be legally authorized[12], but this is unlikely to happen because the exploitation would not be licensed by the environmental authority in the first place.

[12]The concept of groundwater reservoir depletion is somewhat distinct from mineral ore deposit depletion. Depletion of groundwater storage does not mean the complete drainage of the aquifer because is rather unlikely to happen, since an important part of an aquifer is renewable and the stocks can be replenished by recharge. Nevertheless, the continuing abstraction of non-renewable aquifers represents, in fact, reduction in groundwater reserves, triggering several negative side effects. According to Llamas & Custodio, the common undesirable side effects of intensive abstraction of groundwater are the drawdown of the hydraulic head locally and regionally, alteration in the river-aquifer relations (drying up of springs and reduction of river-base flow), alteration in the groundwater quality (normally degradation) and land collapse [23].

The second problem in considering mineral water as a mineral resource refers to the ownership transference of a public asset to a private party. Legally, the miner owns the mined product as part of his private assets. This means that the abstracted mineral water belongs to the miner, differently than users of water resources. As seen before, according to Brazilian Water Law's principles, water rights refer to the private use of a public good with the public good remaining in public hands. Then the mining principle conflicts directly with the constitutional assumption that water is a public good for common use by the people and is inalienable. Under Brazilian Ad- ministrative Law, public assets for common use are not susceptible to being ascribed to individual ownership, since they belong to everyone and the government is in charge of managing them for the benefit of the society [9] . From an economic perspective, the exclusive use of a water resource by a private party depends on an ad- ministrative permit act as stated by a national and regional set of policies and rules. The management of mineral water under the mining exploitation regime seems, therefore, to be incompatible with the premise of the inalie- nability of water.

Finally, some inconsistency between the CM and the CMW rests on the principle of management of mineral water for end use. Mineral water is managed exclusively for bottling process or for spa purposes. Any other use (for example, water supply or irrigation) is not regulated by the CMW and is out of the federal jurisdiction of DNPM. Mining activity, however, is deeply dependent on the fixed location of ore deposits and on its non-renew- ability rather than on the end uses of extracted mineral substances.

CONFLICTS BETWEEN MINERAL AND GROUNDWATER SECTORS

Mineral water management can potentially conflict with other sectors, such as health and sanitation, environ- mental protection, land use planning, and groundwater resource management [24] . Regarding the health sector, there are concurrent administrative levels of the government, such as the Ministry of Health and the Ministry of Mines and Energy, which enact conflicting norms. This is the case for sanitation inspection procedures for min- eral water industries and labelling of bottled mineral water[13]. Another source of conflict lies in the necessary in- tervention in preserved wooded areas (named Permanent Protection Areas defined in the Brazilian Forest Code) to construct the source protection building around mineral water springs, that would be theoretically authorized only under special circumstances defined by law[14]. While environmental law states that Permanent Protected Areas must be preserved, the mineral water law determines deforestation can be carried out in localized portions of those protected wooded areas [24] , which can result in fines by the environmental body.

[13]Ordinance 470/1999 of the Ministry of Mines and Energy conflicts with Resolution RDC 259/2002 of the National Sanitary Surveillance Agency about labelling of bottled products. Likewise, Ordinance 374/2009 of the DNPM conflicts with Ordinance 326/1997 of the Ministry of Health regarding sanitary standards for bottling industries [24].

[14]See Note 10.

Nevertheless, the mineral water sector collides directly and more frequently with groundwater sector. Official statistics demonstrate that almost 190,000 permits for water use had been granted by July 2011, representing a total authorized yield of 6800 m^3/s, where groundwater still represents a minor part, of about 8400 permits and a total yield of 520 m^3/s (or 1,872,000 m^3/h) [25] . Compared with the official statistics for the water resource management sector, the production of mineral water represents a minimal, if not negligible, proportion of the end use of groundwater in Brazil, which is estimated at 1024 exploitation concessions granted by the end of 2011, representing a total yield of 0.28 m^3/s (or 1023 m^3/h) [26] . Although there is such a quantity difference, the potential conflicts between the groundwater and mineral water use still occur, but are rather a localized issue, especially in certain regions where mineral and thermal waters play an important role in production of bottled beverages or for spa tourism.

A well-reported case of conflict takes place in the Caldas Novas region, in the central part of Brazil, an im- portant tourist destination in Goias State, where the presence of thermal water sources has been attracting tour- ists from all over the country for leisure tourism and balneotherapy since the 1960s. A large number of wells (more than 400 wells) had been drilled and groundwater abstraction was almost uncontrolled, resulting in a drastic decrease in the regional water level of the thermal aquifer, with a decrease of more than 50 m from its original level [27] . From 1996, the National Department of Mineral Production (DNPM) has been adopting a set of control measures for the thermal aquifer in the Caldas Novas region, aiming to recover of its falling potentio- metric surface and to maintain the thermal characteristics of the aquifer that was about to be depleted because of groundwater overexploitation [27] . The set of procedures comprised the interdiction of 180 illegal wells, the prohibition of use of thermal water from 155 wells for domestic purposes, the restraint and regular monitoring of the production yield in the 89 authorized thermal water wells and the suspension of granting of new mineral rights in the region[15]. Recently, the control measures successfully led the aquifer to recover about 30 m above the lowest water level registered in the region in the middle 1990s, but some concerns still remain because mon- itoring data have shown a progressive tendency for thermal aquifer drawdown that has not adequately been ad- dressed[16]. On the other hand, the supremacy of thermal groundwater use for mineral water spas has also con- flicted directly with the local public water supply service, because the municipality had their thermal water wells restricted by the federal regulatory body, which ultimately led the matter to be brought before a court of justice to be settled[17]. Only 30 shallow wells are authorized by the State body for abstraction of cold groundwater for domestic uses inside the restricted area [27] .

[15]DNPM Ordinance 52/1999 defined an area of restriction of 107.3 hectares in the Caldas Novas and Rio Quente regions in which 89 min- eral water wells have their yields monitored monthly, with a maximum total yield of 1800 m^3/hour, at a pumping rate of 14 hours/day.

[16]The drawdown tendency of the Araxá Thermal Aquifer System, in Caldas Novas region, can possibly be related to some interference be- tween the thermal water wells authorized by the DNPM and the municipal wells producing groundwater for the public water supply [28] .

[17]According to Andrade & Almeida, the Caldas Novas' municipal water supply service agreed to replace the volume of groundwater ab- stracted from the Araxá Thermal Aquifer System for surface water catchment work carried out in Pirapitinga River. As the work was fin- ished at the end of 1996, the six municipal wells were restricted by the DNPM in 1997. Restrictions of the abstraction of groundwater were later released, from 2008 until the end of 2011 by a judicial sentence, using the justification that the use of surface water by the municipal water service did not adequately supply the local population [28] .

Another reported case of conflict refers to the Nestlé industry in São Lourenço city, a traditional hydromineral resort in Southeast Brazil.

The local population organised protests from 1999 to 2004 against the multinational company, which was suspected of overexploiting the aquifer by intensive pumping of wells for bottled water production [28] . Protests focused on the distrust of the multinational company being granted with mining rights to exploit mineralized groundwater from wells enclosed in a resort with reputed medicinal springs. The supposed- ly high rate of groundwater pumping for bottling purposes would have caused the falling water level of the aquifer, with apparent changes in water composition and the drying of some secular springs. The Nestlé industry was involved in a trial, with great negative repercussions shown both in national and international media. In the end, some irregularities were found in the industry's operation, but there was no direct evidence of aquifer overexploitation due to the production of bottled water. Nestlé had their activities adjusted to the Public Minis- try's requirements by signing an agreement in 2006 that restricted one of the pumping wells, and also limited its total production to 13 million litres of mineral water per year [28] .

Aware of the conflicts of jurisdiction between the federal regulatory body (management of mineral water) and State institutions (management of water resources), the National Water Resource Council (CNRH) in 2002 started a long discussion on this issue, conducted within its Groundwater Technical Chamber (CTAS). The dis- cussion about the integration between the management of mineral water and groundwater were particularly lively in 2004, when a seminar was held focusing conceptual and legal issues in each sector. The discussion was somewhat heated and opinions were clearly polarized into two divergent points of view: on the one side repre- sentatives of the National Department of Mineral Production (DNPM) and of the National Confederation of In- dustry were against any attempt to integrate mineral water developments with the permit system of the National System of Water Resource Management (SINGREH). On the other side, representatives of Federal and State bodies for the environment and water resource management argued the opposite [16] [29] [30] . In the end, the Council published Resolution 76/2007 focusing mainly on guidelines to promote the dialogue between federal and states administrative institutions, as well as the sharing of technical information of each permit system. This resolution seeks, to stimulate the "integration and coordinated action" of the organisations involved (federal and state agencies) without formally stating that SINGREH should be an additional management instrument for mineral water developments[18].

[18]Although there was no consensus in the CNRH, Serra observes that some states actually provided legal means for the integration of mineral water claims into the SINGREH permit system (states of Bahia, Paraná, Pernambuco and Rio Grande do Norte). This occurs because States have legislative authority to enact administrative laws regarding environmental protection and water use permits, which vary from State to State [15].

A likely reason for this sort of conflict is that mineral water today is not so far distinct from "ordinary" groundwater. According to Queiroz, 88% of all mineral water sources in Brazil produce very lightly to lightly mineralized water, showing a total concentration of dissolved chemical compounds no higher than 300 mg/L [31] . About 10% of the sources are classified as potable table water and 27% of the sources are classified as mineral water exclusively due to insignificant amounts of fluoride as a noteworthy component [31] . This study also shows that 43% of all Brazilian mineral water sources are characterized exclusively by intermittent physical and chemical parameters, such as temperature and/or temporary radioactivity, which means they are truly min- eral waters only at, or near to, their sources. Away from their sources, they are only potable table waters (or common potable groundwater) because their temperature has changed and their dissolved gases have been lost at the time they are available for consumption. While those intermittent physical-chemical characteristics are sig- nificant for spa purposes, they are irrelevant for the consumer of bottled mineral water.

Only a small part of all mineral water sources in Brazil, about 20%, is characterized by greater diversity of chemical compounds and/or high temperature and/or natural carbonation, characteristics that are strictly in ac- cordance with the legal chemical limits established in Article 35 of the CMW and with the legally stated defini- tion of a supposed medicinal action for mineral water. This percentage is roughly the proportion of mineral wa- ter sources (14%) that historically were devoted to some medicinal purpose or aimed at leisure tourism in spas and hydromineral resorts [31] .

Considering only bottled mineral water, one third are very lightly mineralized (total dissolved solids < 50 mg/L), associated with fast water circulation through shallow aquifers (pH < 6, acidic), theoretically highly vul- nerable to surface-driven contamination, and furthermore one quarter of the sources shows some signs of anth- ropic impact indicated by the presence of nitrate[19], even though they are potable [32] . There are some sources in the North and Northeast of Brazil that produce mineral water compositionally similar to rainwater, as they show total dissolved solid contents below 5 mg/L and pH between 4 and 5 [17] [32] .

[19]Bertolo et al. argue that groundwater with nitrate concentration above 3 mg/L has been somewhat impacted by surface derived anthropic contamination [32].

In fact, the CMW currently fails to sort out conflicts between the water and mining sectors because the law does not provide reasonable well-defined limits, based on intrinsic parameters, which can without doubt differ- entiate mineral water, potable table water and common groundwater. With some exceptions, in practice, they are almost the same, because the CMW allows a loose interpretation of the legal parameters used to classify any potable groundwater either as mineral water or as potable table water.

In the 1940s, when the CMW was drafted, there was no stated policy at federal level to deal with other eco- nomic uses of groundwater apart those regulated by the CMW. Furthermore, the limited knowledge in Hydroge- ology in the 1940s and the lack of technology in underground works have possibly led to the misconception that mineralized groundwater was disconnected from surface waters, causing mineral water management to be regu- lated more closely to the mining sector, completely separated from surface water resource management [15] . The CMW's provisions worked without major conflicts for years, until the Federal Constitution came into force in 1988, which introduced an environmental dimension for water, previewed a national-level integrated water resource management, and empowered the States with jurisdiction over groundwater.

With groundwater being defined by the Constitution as a natural resource under States' jurisdiction, there is a risk that potable table water would in fact be interpreted as being "common" groundwater, since they are technically the same. If this were the case, the CMW's provisions designating potable table water as a mineral re- source would conflict directly with the 1988 Federal Constitution. Accordingly, neither potable table water nor mineral waters would be regarded as assets belonging to the Union, because the Federal Constitution makes no distinction between different types of water, but includes all groundwater under the States' domain. As vehe- mently argued by Serra, if no exception is made for groundwater, the CMW as a whole would be implicitly re- voked by the 1988 Federal Constitution and mineral water would fall under States' competence as a water re- source [15] . This is a controversial point of view, however, and is not shared by the Ministry of Mines and Energy's advisory board of Federal Attorneys (Advogado Geral da União). They argue that the CMW is in ac- cordance with the Federal Constitution, because mineralized groundwater is solidly defined as a mineral re- source at infra-constitutional level [29] .

THE NEED FOR UPDATING THE BRAZILIAN LEGISLATION FOR MINERAL WATER

The concepts used by the CMW for defining mineral water lie in the sparse scientific and medical knowledge about the therapeutic uses of mineralized, radioactive and thermal waters that was available in the 1940s. Some properties of the water that are directly linked to the source, such as high temperature and dissolved gas content, are relevant for therapeutic purposes and have been extensively reported as beneficial to health in both recent and older international literature on spa therapy [33] -[35] . Some therapeutic effects of mineral waters, however, are currently asserted in law, but are unsubstantiated by science. The actions of chemicals are now much better understood than they were when the law was drafted, and it is now clear that the law defining the therapeutic benefits of some of the chemical compounds found in mineral water seems to be misleading.

For example, nitrate and arsenic are defined by the CMW as likely to be associated with some medicinal ac- tion, but they in fact have been found to have some harmful effects in humans, so that their concentration in drinking water is now determined by rigid standards[20]. On the other hand, the presence of radioactivity (perma- nent or temporary) in water, which for a long time Crenotherapy had credit with the mild therapeutic effects of lightly mineralized water [36] , has had its beneficial action on health questioned, and it has even been described as hazardous to humans [17] [37] . If there are controversies about the benefits or possible harms to humans re- lated to certain components in drinking water, precautionary concerns should be taken into account in using these parameters for mineral water classification and its use for human consumption.

[20]According to Resolution 274/2005 of the National Health Surveillance Agency, the maximum amount of dissolved nitrate and arsenic in drinking water is, respectively, 50 mg/L and 0.01 mg/L.

The concept of mineral water carrying some medicinal effect is currently in disuse in Brazil, mainly because medicinal practices using mineral and thermal waters are virtually extinct due to the lack of experts in Creno- therapy [14] [15] . As the linkage between medicinal therapy and mineral water has been lost, the principles that underpin the Brazilian legal framework for dealing with mineralized groundwater should be revised. As pointed out by the reputed Brazilian jurist, Miguel Reale, "the legal rule must be formally valid and socially effective" [38] . In other words, the effectiveness of the law is dictated by its engagement with the social experience, to the point that "the disuse [of a law] can occur because the legal rule has never been applied or, at a certain time, has ceased to be applied, since the obedience to a diverse customary norm started to prevail within the community, with the oblivion of the legal rule" [38] . The same author also argues that: "the public authorities have a duty to avoid the divorce between social reality and certain legal rules, which are not, or never have been effective, be- cause it [the law] is in conflict with trends and authentic dominant interests within the community" [39] .

The current political climate in Brazil is particularly favorable to making public authorities aware of the ur- gent need for reopening the discussion for updating mineral water legislation, because the government has been making efforts to modernize outdated laws. In this direction, the federal government has released the bill that will replace the 1967 Code of Mining and will improve the institutional and legal framework of the mining ac- tivity in Brazil. The bill number 5.807/2013 is before the National Congress since June of 2012 and several changes in principles and management instruments for dealing with mineral resources are likely to happen by the end of 2014. The expected changes, however, will not affect the current management style of mineral waters.

A desirable scenario would be for the government to open a public consultation process followed by a trans- parent discussion through public meetings, involving different sectors of society with competing interests in the mineral water issue, such as the mining, groundwater, health and food sectors. Academic knowledge and the experience of other countries would be of great value in supporting the discussion and the law-making process. The government should provide an open-minded debate for answering the big question: which is the model for mineral water management that society really wants and should pursue? The multidisciplinary debate should lead to answers regarding the establishment of (new) principles and concepts for mineral water, the definition of the management instrument, a decision about the legal regime of exploitation and the determination of well-defin- ed technical parameters that unquestionably differentiate mineral water from normal groundwater.

The current design of the mineral water sector has evolved so that most of the mineral water sources today are barely distinguishable from normal groundwater. The connection between mineral water and medicine has been lost for a long time and so is the linkage with the principles that underpin the mining legal framework. Keeping mineral water management in the mining sector can only be justified if the principles governing mineral water legislation are in accordance with the non-renewability of aquifers and with the fixed location of the mineral water sources[21]. Otherwise, mineral water will continue to be treated like an alien substance with an alien model of management within the mining sector.

[21]That would be the case of mineralized groundwater from fossil aquifers, for example.

Nevertheless, keeping mineral water management in the mining sector does not sort out concerns about the compatibility of the mining legal regime for mineral water exploitation, because the mining alienability principle conflicts directly with the environmental dimension of groundwater.

CONCLUSIONS

For reasons that have little to do with hydrogeology, but much to do with currently disused traditional spa prac- tices, mineral water in Brazil is considered to be a mineral substance and even today it is treated differently to water resources. The CMW is a curious case in which the law did promote a change in the nature of a natural resource, namely, groundwater. The legal ascription of mineral water as a mineral substance should never ex- clude its natural condition as a groundwater resource. In fact, it is almost physically impossible to separate mi- neralized groundwater from "ordinary" groundwater because they are interconnected in an aquifer, taking part of a dynamic system composed of flowing water and rock. Furthermore, mineral water has now acquired a water resource status in its own right, taking into account how legal instruments for water management have evolved in Brazil in the last decade.

The cases of conflict reported in this article highlight the inversion of priorities stated by the Constitution, which was supposed to have the supremacy of public over private interests. In the Central Brazil conflict case, spa tourism is favoured in the end use of mineralized water for private interests rather than for the public water supply service. The regime of mineral water production under the mining legal framework sometimes denies some principles and guidelines of environmental and water resource legislations that should not happen today.

There will be inevitable conflicts of jurisdiction on mineralized groundwater while the CMW remains in force, because it is founded on outdated definitions and disused concepts regarding water intrinsic parameters and end uses. The CMW has ceased to be effective and barely reflects the current design of the mineral water market, since it has been drafted since the 1960s primarily for the production of bottled beverages rather than for medi- cinal purposes. Bottled mineral water today favours characteristics agreeable to consumer tastes (lightly minera- lized bottled water), with the oblivion or the principles originally stated by law (mineral water as a remedy). More than 80% of mineral water sources in Brazil are currently devoted to production of bottled water. For the consumers of bottled mineral water, the only thing that really matters is the quality of groundwater for drinking and not its presumable medicinal effects on humans.

There is an urgent need to reopen the discussion aimed at a possible revision of the CMW. This piece of leg- islation has been frozen in time and does not adequately take into account the deep changes in the way society has been dealing with water, in general, and with the environment, in particular. For the discussion process, some aspects should be taken into account, such as serious concerns about the compatibility of the CMW's prin- ciples with the environmental dimension of groundwater. The ascription of mineral water as a mineral substance belonging to the Union's assets and the adoption of the legal regime of mining are frankly in disagreement with the constitutional premises of groundwater under the States' jurisdiction and with the inalienability of water as a public good for common use.

ACKNOWLEDGEMENTS

The first author thanks the DNPM for supporting this research and also the CNPq (Brazilian Council of Technological and Scientific Development) for the post-doctoral scholarship provided during 2012-2013 period.

REFERENCES

1. Porto, M. and Kelman, J. (2000) Water Resources Policy in Brazil. Rivers—Studies in the Science Environmental Policy and Law of Instream Flow, 7, 250-259.http://www.kelman.com.br/pdf/Water_Resources_Policy_In_Brazil_2.pdf

2. Benjamin, A.H., Marques, C.L. and Tinker, C. (2005) The Water Giant Awakes: An Overview of Water Law in Brazil. Texas Law Review, 83, 2185-2244.

3. ANA, Brazilian National Water Agency (2007) GEO Brazil Tematic Series, Water Resources Component of a Series of Reports on the Status and Prospects for the Environment in Brazil. ANA/PNUMA, Brasília.http://arquivos.ana.gov.br/institucional/sge/CEDOC/Catalogo/2010/GEOBrasilResumoExecutivo_Ingles.pdf

4. Hirata, R., Zoby, J.L.G. and Oliveira, F.R. (2010) Groundwater: Strategic or Emergency Reserve. In: Bicudo, C.E., Tundisi, J.G. and Scheuenstuhl, M.C.B., Org., Waters of Brazil: Strategic Analysis, Instituto de Botânica, São Paulo, 149-161. (In Portuguese)http://www.ianas.org/books/aguas_do_brasil_Final_02_opt.pdf

5. ANA, Brazilian National Water Agency (2011) Permit System for Water Resources Use. ANA, Brasília. (Cadernos de

Capacitação em Recursos Hídricos, Vol. 6) (In Portuguese)http://arquivos.ana.gov.br/institucional/sge/CEDOC/Catalogo/2012/OutorgaDeDireitoDeUsoDeRecursosHidricos.pdf

6.   Drummond, J. and Barros-Platiau, A.F. (2006) Brazilian Environmental Laws and Policies, 1934-2002: A Critical Overview. Law & Policy, 28, 83-108.http://dx.doi.org/10.1111/j.1467-9930.2005.00218.x

7.   Daibert, A. (2009) Historical Views on Environment and Environmental Law in Brazil. The George Washington International Law Review, 40, 779-840.

8.   McAllister, L.K. (2008) Revisiting a "Promising Institution": Public Law Litigation in the Civil Law World. Georgia StateUniversity Law Review, 24, 693-734.

9.   Pompeu, C.T. (2010) Water Rights in Brazil. 2nd Edition, Editora Revista dos Tribunais, São Paulo. (In Portuguese)

10.   Di Pietro, M.S.Z. (2010) Private Use of a Public Good by a Private Party. 2nd Rdition, Editora Atlas S/A, São Paulo. (In Portuguese)

11.   Freire, W. (2010) Fundamentals of Mining Law. Jurídica Editora, Belo Horizonte.

12.   Trindade, A.D.C. (2009) Mining Law Principles. In: Souza, M.M.G., Coord., Mining Law in Evolution, Editora Mandamentos, Belo Horizonte, 47-76. (In Portuguese)

13.   Trindade, A.D.C. (2011) Perspectives for a Mining Law Reform. In: Martins, J., Lima, P.C.R., Queiroz Filho, A.P., Schüller, L.C. and Pontes, R.C.M., Coord., Mineral Sector: On Route to the New Legal Framework, Câmara dos Deputados, Brasília, Série Cadernos de Altos Estudos 8, 205-220. (In Portuguese) http://www2.camara.leg.br/a-camara/altosestudos/arquivos/setor-mineral-rumo-a-um-novo-marco-legal/setor-mineral-rumo-a-um-novo-marco-legal

14.   Quintela, M.M. (2004) Thermal Knowledge and Practices: A Compared Perspective in Portugal (São Pedro do Sul Thermal Spa) and in Brazil (Caldas da Imperatriz Hot Springs). História, Ciências, Saúde, 11, 239-260. (In Portuguese)http://www.scielo.br/scielo.php?script=sci_pdf&pid=S0104-59702004000400012&lng=en&nrm=iso&tlng=pt

15. Serra, S.H. (2009) Mineral Waters of Brazil. Millennium Editora, Campinas.

16. Caetano, L.C. (2005) Mineral Water Policy: An Integrated Proposal for the Rio de Janeiro State. Ph.D. Thesis, State University of Campinas, Campinas.

17. Bertolo, R.A. (2006) Reflections on the Mineral Water Classification and Chemical Characteristics of Bottled Mineral Water in Brazil. Proceedings of the 14th Brazilian Groundwater Congress, Curitiba, 7-18 November. (Revista Águas Subterrâneas, Supplementary Issue) (In Portuguese)http://aguassubterraneas. abas.org/asubterraneas/article/view/23114/15229

18. Nickel, E.H. (1995) The Definition of a Mineral. The Canadian Mineralogist, 33, 689-690.

19. Rebouças, A.C. (2006) Groudwaters. In: Rebouças, A.C., Braga, B. and Tundisi, J.G., Org., Freshwaters in Brazil, 3rd Edition, Editora Escrituras, São Paulo, 111-144. (In Portuguese)

20. Margat, J., Foster, S. and Droubi, A. (2006) Concept and Importance of Non-Renewable Resources. In: Foster, S. and Louks, D.P., Eds., Non-Renewable Groundwater Resources, A Guidebook on Socially-Sustainable Management for Water-Policy Makers, UNESCO, IHP-VI, Series on Groundwater 10, 13-24. (IHP/2006/GW-10)http://unesdoc.unesco.org/ images/0014/001469/146997e.pdf

21. Foster, S., Nanni, M., Kemper, K., Garduño, H. and Tuinhof, A. (2005) Utilization of Non-Renewable Groundwater: A Socially-Sustainable Approach to Resource Management. The World Bank, Washington DC. (GWMTE Briefing Note Series, Note 11). http://documents.worldbank.org/curated/en/2003/01/5161421/ utilization-non-renewable-groundwater-socially-sustainable-approach-resource-management

22. Custodio, E. (2002) Aquifer Overexploitation: What Does It Mean? Hydrogeology Journal, 10, 254-277. http://dx.doi.org/10.1007/ s10040-002-0188-6

23. Llamas, M.R. and Custodio, E. (2003) Intensive Use of Groundwater: A New Situation Which Demands Proactive Action. In: Llamas, M.R. and Custodio, E., Ed., Intensive Use of Groundwater, Challenges and Opportunities, A. A. Balkema Publishers, Lisse, 13-34.

24. Caetano, L.C., Pereira, S.Y. and Dourado, F. (2012) Conflicts of Mineral Water Management in Brazil—Study Case: Rio de Janeiro State. Holos Environment, 12, 132-146. (In Portuguese) http://www.periodicos.rc.biblioteca.unesp.br/index.php/holos/article/view/2080/4935

25. ANA, Brazilian National Water Agency (2012) Water Resources Conjuncture in Brazil. 2012 Report. ANA, Brasília. (In Portuguese) http://arquivos.ana.gov.br/imprensa/arquivos/Conjuntura2012.pdf

26. DNPM, National Department of Mineral Production (2012) 2012 Mineral Summary. DNPM, Brasília, Vol. 32, 3-4. (In Portuguese) https://sistemas.dnpm.gov.br/publicacao/mostra_imagem.asp?IDBancoArquivoArquivo=7368

27. Andrade, A.M.A. and Almeida, L. (2012) Potenciometric Level Performance of the Thermal Caldas Novas Aquifer ? Goias State and Control Measures Imposed by the National Department of Mineral Production (DNPM). Águas Sub- terrâneas, 26, 99-112. (In Portuguese) http://aguassubterraneas.abas.org/asubterraneas/article/view/25048/17678

28. Guimarães, B.C. (2009) Environmental Collective Law and the (un)Sustainable Exploitation of Mineral Waters. Mandamentos, Belo Horizonte. (In Portuguese)

29. Scalon, M.G.B. (2011) Mineral Waters and Water Resources: A Perspective of Integrated Management. Revista de Direito, Estado e Recursos Minerais, 1, 131-160. (In Portuguese)http://seer.bce.unb.br/index.php/rdern/article/view/5173

30. Reis, A.M. (2005) Searching for an Integrated Management of Mineral Waters in Brazilian Legislation. In: Freitas, W.P., Ed., Environmental Law in Evolution, Vol. 4, Editora Juruá, Curitiba, 15-39. (In Portuguese)

31. Queiroz, E.T. (2004) Mineral Waters of Brazil: Distribution, Classification and Economic Importance. DNPM, Brasília. (In   Portuguese)http://www.dnpm.gov.br/mostra_arquivo.asp?IDBancoArquivoArquivo=377

32. Bertolo, R., Hirata, R. and Fernandes, A. (2007) Hydrogeochemistry of Bottled Mineral Waters in Brazil. Revista Brasileira de Geociências, 37, 515-529. (In Portuguese)http://rbg.sbgeo.org.br/index.php/rbg/article/view/A-1647/984

33. Lopes, R.S. (1956) Mineral Waters of Brazil, Value and Therapeutic Uses. Ministério da Agricultura, Rio de Janeiro. (In Portuguese)

34. Mourão, B.M. (1992) Hydrological Medicine—Modern Therapy of Mineral Waters and Healing Centres. Secretaria Municipal de Educação, Poços de Caldas. (In Portuguese)

35. Bender, T., Karagülle, Z., Bálint, G.P., Gutenbrunner, C., Bálint, P.V. and Sukenik, S. (2005) Hydrotherapy, Balneotherapy, and Spa Treatment in Pain Management. Rheumatology International, 25, 220-224. http://dx.doi.org/10.1007/s00296-004-0487-4

36. Nasermoaddeli, A. and Kagamimori, S. (2005) Balneotherapy in Medicine: A Review. Environmental Health and Preventive Medicine, 10, 171-179. http://dx.doi.org/10.1007/BF02897707

37. WHO, World Health Organization (2008) Guidelines for Drinking-Water Quality. http://www.who.int/entity/water_sanitation_health/dwq/GDW9rev1and2.pdf

38. Reale, M. (2001) Preliminary Lessons of Law. 25th Edition, Saraiva, São Paulo. (In Portuguese)

39. Reale, M. (2002) Law Philosophy. 19th Edition, Saraiva, São Paulo. (In Portuguese)

Chapter 7

Bioaccumulation of Heavy Metals in Contaminated River Water-Uppanar, Cuddalore, South East Coast of India

Usha Damodharan[1]

[1]Department of Ecology and Environmental Sciences, Pondicherry Central University, Pondicherry, India

INTRODUCTION

Industrialization and human activities have partially or totally turned our environment into dumping sites for waste materials. As a result, many water resources have been rendered polluted and hazardous to man and other living systems [1]. The toxic substances discharged into water bodies are not only accumulated through the food chain [2], but may also either limit the number of species or produce dense populations of microorganisms [3]. Aquatic ecosystems are affected by several stresses that significantly weaken biodiversity. River pollution is an environmental problem in the world. They are subjected to various natural processes such as the hydrological cycle occurring in the environment, Because of unprecedented development, human beings are responsible for choking several aquatic ecosystems to death. Storm water runoff and carry out of sewage into rivers are two common ways that various nutrients and other pollutants enter the aquatic ecosystems resulting in pollution [4, 5]. Heavy metal contamination particularly the non-essential elements may have distressing effects on the ecological balance of the recipient aquatic environment with a diverse of organisms including fish. It has particular significance in ecotoxicology, since the heavy metals are highly persistent and have the potential to bio accumulate and bio magnify in food chain, and become toxic to living organisms at higher trophic levels in nature.

Uppanar River is considered to be one of the highly polluted rivers in south east coast of India due to industrialization. SIPCOT (Small Industrial Promotion Corporation of Tamil Nadu, covering an area of about 520 acres with 52 industries) is located on the bank of the Uppanar River at Cuddalore. It was established for chemical, petrochemical, pharmaceutical, biocides, fertilizer, fungicides, chlor–alkai and metal processing industries etc. Many possible environmental contaminants could be discussed in a review of toxic substances from industrial sources. The combined effect of all these might be the reason of frequent fish death and depletion of aquatic ecosystems in this area. Indiscriminate discharge of partially treated effluents from SIPCOT industrial complex into coastal environment affects both biotic and abiotic system and finally causes some ill effects to human beings through food chain. A detailed study was made on the bio accumulation of heavy metals in the food chain and the spectrum of issues and consequences were discussed in the present study.

The specific objectives of this study are:
- The interpretation of the impact of pharmaceutical industrial effluent on surface water quality of Uppanar River in Cuddalore.
- Find out the suitability of river water for fisheries.
- To evaluate the health risk of fish consumption by human beings collected from the river.

Study Area

Uppanar River is a stream in Cuddalore (Lat.11/ 43'N, Long. 79/ 46' E) (Fig. 1). It flows between Cuddalore and Chidambaram Taluks and joins with the Bay of Bengal by the mouth of Gadilam River. It runs behind the SIPCOT (State Industrial Promotion Corporation of Tamil Nadu Limited) industrial complex which consists of Pharmaceutical industries, fertilizers, dyes, chemicals, mineral processing plants and metal based industries. The river receives the partially treated and untreated effluents of these industries through small channels and pipeline. The water at lower reaches is polluted more when compared to the upper reaches. In addition to the industrial wastes, the river also receives the municipal wastes and domestic sewage from the Cuddalore old town. As the river receives the treated and partially treated effluents from nearly 55 industries, it is said to be highly polluted. Pharmaceutical industrial effluent before and after treatment and four stations in the Uppanar River, the river (Outfall), Upstream (uncontaminated -Semmankuppam) and Downstream (Contaminated-Kudikadu) were selected for the heavy metal (Cu, Cd, Mn, Zn and Pb) analyses and seasonal variations were reported. A study on the transfer of Cd, Pb, Cu, mn and Zn through the food chain of the river Uppanar proved the existence of bioaccumulation in fish.

Sampling Methods and Sample Preparation

Water samples were collected with 1 L polyethylene bottles which were previously cleaned by washing with non-ionic detergent, rinsed with tap water and later soaked in 10% HNO_3 for 24 hours and finally rinsed with deionized water prior to use. During sampling, sample bottles were rinsed with sampled water three times and then filled to the brim at a depth of one meter below the wastewater from each of the four designated sampling points. Temperature and pH measured immediately after collection. The waste water and river water samples were digested and heavy metals were determined using Atomic Absorption Spectrophotometer (AAS) as described in the APHA standard methods (1992).

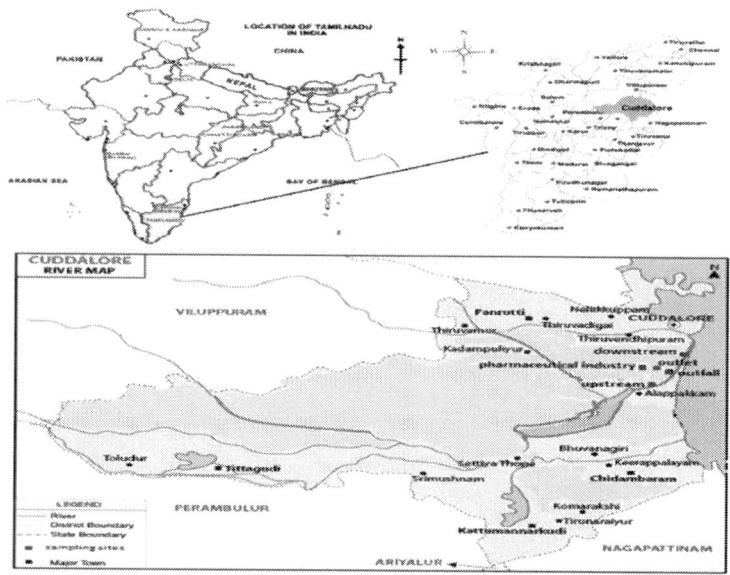

Figure 1: The map of Cuddalore and Uppanar River sampling sites.

Dried samples of muscle tissues from each fish were digested using microwave digestion system. After digestion, the residues were diluted to 25ml with 2.5% of HNO_3. The Instrument (Atomic Absorption Spectrophotometer – AAS) was calibrated with standard solutions and prepared from commercial materials. The water used was deionized and distilled. The metal analysis of the tissue and water samples (Cd, Cu, Mn, Zn, Pb) were carried out by using Atomic Absorption Spectrophotometer (AAS).

Result and Discussion

The spatial variation of the heavy metals, Cd, Cu, Pb, Zn and Mn) along the six sampling locations of all the four seasons was shown in Fig 2. The average mean concentrations of Cd, Cu, Pb, Zn and Mn were higher in the untreated effluents than the treated effluents.

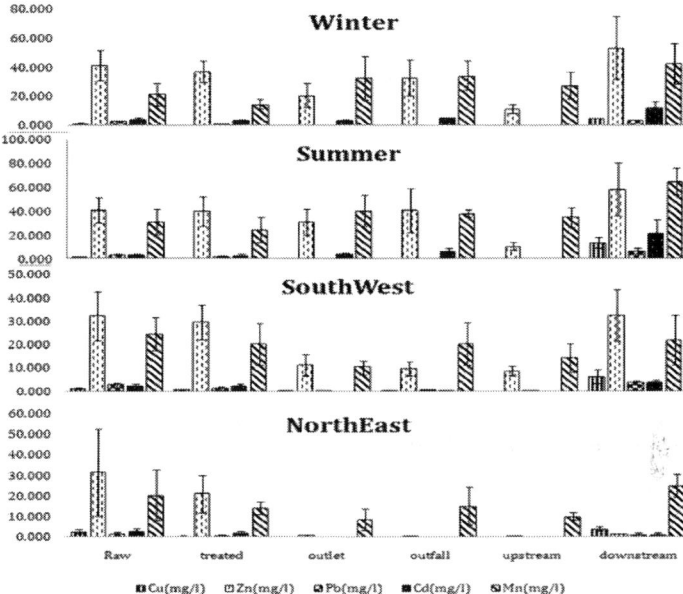

Figure 2: Mean and S.D of heavy metal (Cu, Pb, Cd, Mn and Zn) concentrations in pharmaceutical industry effluents(Raw & Treated) and Uppanar River water at four sampling sites.

Cadmium

Cadmium concentrations in unpolluted natural waters are usually below 1 μg/l. Contamination of drinking-water may occur as a result of the presence of cadmium as an impurity in the zinc of galvanized pipes or cadmium-containing solders in fittings, water heaters, water coolers and taps. Drinking-water from shallow wells of areas in Sweden where the soil had been acidified contained concentrations of cadmium approaching 5 μg/l. In Saudi Arabia, mean concentrations of 1– 26 μg/l were found in samples of potable water, some of which were taken from private wells or cold corroded pipes (Mustafa et al., 1988). Levels of cadmium could be higher in areas supplied with soft water of low pH, as this would tend to be more corrosive in plumbing systems containing cadmium. In the Netherlands, in a survey of 256 drinking-water plants in 1982, cadmium (0.1–0.2 μg/l) was detected in only 1% of the drinking-water samples. Cadmium has been shown to induce carcinogenesis by both the inhalation and parental routes of exposure. The variations in the heavy metal concentration of both the untreated and treated effluents were due to the heavy metal decreasing efficiency of primary and secondary treatment of the Effluent Treatment Plant (ETP). The average mean concentration of Cd at the river water ranged from (0.011 – 21.213 ppm) during all the four seasons. There was a higher fluctuation in the various sampling sites of river, which was attributed to the reason of the other industrial and anthropogenic sources. Upstream Cd concentration during all the four seasons were several times lesser than the Cd concentration of raw effluent, which indicated the dilution effect of the river weather and also it showed that this site is free from anthropogenic inputs, and only the natural effects are the predominant factors in this particular site.

Copper

Copper is a natural element which is widely distributed in soils, rocks and in rivers and the sea. The Cu is widely used in society and yet is potentially quite toxic to life in rivers. The present result of average mean concentration of Cu at the various sampling stations of river water ranged from 0.230 – 13.313mg/l during all the four seasons. The Cu concentration at outfall increased two folds compared to the outlet that could be attributed to the reason of anthropogenic activities, agriculture runoff, sludge from publicly-owned treatment works (POTWs) and municipal and industrial solid waste dumped into the river water. Copper is released in to water as a result of natural weathering of soil and discharges from industries and sewage treatment plants [7, 8]. The Cu concentration in downstream were several folds higher than the raw effluent Cu concentration. It may be attributed to domestic sewage and run-off from extensive farmed areas [9]. Copper compounds which are used in electroplating industries such as cupric sulphate and cupric acetate and in fertilizers such as copper naphthenate and paint industries such as cuprous oxide, Ceramics and glass industries such as cupric acetate, cuprous and cupric oxides used as pigments and for making glazes were discharged through the treated industrial effluents. Other than this copper released through the domestic activities such as human wastes flushed through the toilets, washing and bathing water etc. Copper occurs naturally in all foods and water and, in small concentrations, plays an essential role in the human diet. Copper in the dissolved form is potentially very toxic to aquatic animals and plants, especially to young life-stages such as fish larvae. The toxicity of copper is however greatly reduced when it is bound to particulate matter in the river water and when the water is hard. The industries and public should recognize the need to monitor the concentrations in discharges and in rivers closely, to ensure that Water Quality Objectives are not exceeded.

Lead

Exposure to lead causes a variety of health effects, and affects children in particular. Water is rarely an important source of lead exposure except where lead pipes, for instance in old buildings, are common. Removal of old pipes is costly but the most effective measure to reduce lead exposure from water. The higher concentration of Pb at various sampling sites of river water could be attributed to the reason of less soluble of Pb containing minerals in natural water and its concentration diluted through the dilution effect of the water [10, 11]. The Pb concentration in downstream of all the four seasons was several folds higher than the raw effluents. The profile of the Pb showed that it did not have only one source; furthermore higher concentration of Pb in the downstream indicated the presence of contamination through various industrial effluents of SIPCOT area and local anthropogenic inputs. Lead is rarely found in source water, but enters tap water through corrosion of plumbing materials. The most common problem is with brass or chrome-plated brass faucets and fixtures which can leach significant amounts of lead into the water, especially hot water. Most industrially processed lead is applied for fabricating computer and TV screens. The lead compound tetra-ethyl lead is applied as an additive in fuels. This organic lead compounds is quickly converted to inorganic lead, and ends up in water, sometimes even in drinking water. Fortunately, this form of release of lead is less and less abundant. Lead accumulates in leg tissue. The most severe type of lead poisoning causes encephalopathy. Lead toxicity is induced by lead ions reacting with free sulfydryl groups of proteins, such as enzymes. These are deactivated. Furthermore, lead may interact with other metal ions.

Zinc

Zinc can be introduced into water naturally by erosion of minerals from rocks and soil; however since zinc ores are only slightly soluble in water. Values of 5–22 mg have been reported in studies on the average daily intake of zinc in different areas. The zinc content of typical mixed diets of North American adults varies between 10 and 15 mg/ day. Drinking-water usually makes a negligible contribution to zinc intake unless high concentrations of zinc occur as a result of corrosion of piping and fittings. Under certain circumstances, tap water can provide up to 10% of the daily intake (WHO). High natural levels of zinc in water are usually associated with higher concentrations of other metals such as lead and cadmium. Mostly, the zinc is introduced into water by artificial pathways such as by-products of steel production or coal-fired power stations, or from the burning of waste materials. Zinc is also used in some fertilizers that may leach into groundwater. Older galvanized metal pipes and well cribbings were coated with zinc that may be dissolved by soft, acidic waters. Zinc is an essential nutrient for body growth and development; however drinking water containing high levels of zinc can lead to stomach cramps, nausea and vomiting. Although the Zn is normally found little amount in nature, It is also emitted through effluents of many commercial industries during mining and smelting (metal processing) activities. In the present study upstream Zn concentration during all the four seasons was several times lesser than the Zn concentration of raw effluent, which showed that there was no adverse effect of effluent on the upstream. The Zn concentration in downstream ranged from 1.130 – 58.046 mg/L of all the four seasons which was attributed to the greatest frequency of near source areas like hazardous waste sites and the release of industrial effluents through the transmission of iron pipes. Urban runoff, mine drainage, and municipal sewages are the more concentrated sources of zinc in water.

Manganese

The element manganese is present in over 100 common salts and mineral complexes that are widely distributed in rocks, in soils and on the floors of lakes and oceans. Industrial emissions containing manganese oxides are the principal source of manganese in the atmosphere. The total atmospheric emission of manganese from anthropogenic sources in India was estimated to be 1225 t in 1984; 78.5% of this originated from industrial processes, mainly related to metal alloy production. Emissions stemming from gasoline-powered motor vehicles accounted for a further 17.2%, whereas the remaining 4.3% of atmospheric manganese emissions were due to the burning of coal for power generation, solid waste incineration and pesticide application. In the present study the Mn concentration in downstream ranged from 21.736 – 64.837 mg/L of all the four seasons and it was several folds higher than the raw effluent, which could be attributed to the reason of its usage in the cleaning, bleaching, manufacturing of iron, steel alloys, batteries, glass and fireworks industries and improper discharge of the effluents from these industries [12,13]. Manganese is an essential element in humans and animals, functioning both as an enzyme co-factor and as a constituent of metalloenzymes. Gross deficiencies of manganese have never been observed in the general population, but a recent experimental study involving human subjects fed a manganese-deficient diet (0.11 mg/d) resulted in the development of dermatitis and hypocholesterolaemia and elevated concentrations of serum calcium and phosphorus. A statistical analysis of the metabolic studies showed that a daily manganese intake of approximately 5 mg is required to consistently maintain a positive balance.

The present results showed that the metal concentration decreased in the ranking order of Zn>Mn>Cu>Cd> Pb during all the four seasons. According to the seasons, it was in the following ranking pattern summer>winter>Southwest>Northeast.

Most of the dissolved heavy metals showed high concentrations during the summer period than that of the other seasons. ANOVA of the results showed that the metal concentrations were significantly different between sampling stations and the four seasons. The highest flows occurred during the northeast monsoon from October to December. It was also observed that for all (five) metals studied; there was a trend of increasing concentrations from the upstream stations to the downstream stations.

BIO CONCENTRATION FACTOR

A study on the transfer of Cd, Pb, Cu, Mn and Zn through the food chain of the river Uppanar proved the existence of bioaccumulation in fish. The Bio- concentration factor of the heavy metals in muscle tissues of fish from the river water are presented in Table 1. The BCF of Cu ranged from 0.001 to 0.009, that of Pb ranged from 0.061-0.100, Cd from 0.003-0.004, Zn from 0.031-0.083, and Mn 0.408-0.922.The BCF value of Cu was found highest (0.009) in *L. calcarifer* and it was low in *A. thalassinus* (0.001), while the BCF value for Pb was highest (0.100) in *M. cephalus* and was lowest in *L. calcarifer* (0.061). The BCF value of Cd was highest (0.004) in *T. mossambica* and *M. cephalus*(0.004) and it was low 0.003 in *A. thalassinus* and *L. calcarifer*. The BCF value of Zn was highest (0.083) in *A. thalassinus* and was lower (0.031) in *L. calcarifer*. The BCF value of Mn was highest (0.922) in *T. mossambica* was highest and was lower (0.408) in *L. calcarifer*. The trends of BCF for heavy metals in four species of fish were in the ranking order of Mn>Pb>Zn>Cu>Cd.

Table 1: Bio concentration factor of heavy metals in muscle tissues of fish from the river water

BCF	L.alcarifer	M.cephalus	A.thalasinuss	T.mossambica
Cu muscles/Cu water	0.009	0.002	0.001	0.002
Pb muscles/Pb water	0.061	0.100	0.063	0.074
Cd muscles /Cd water	0.003	0.004	0.003	0.004
Zn muscles/Zn water	0.031	0.041	0.083	0.073
Mn muscles/ Mn water	0.408	0.798	0.707	0.922

[i] - BCF =Conc. in muscle tissues of fish (dry weight basis) / Conc. in water of the river

Due to the improper discharge of partially treated industrial waste water and domestic sewage water into the Uppanar river, the level of heavy metals recorded in water of the downstream and the heavy metals other than Zn and Mn in the water of upstream were generally high and their values exceeded the maximum permissible limit when compared with the limits of WHO (Pb – 0.05, Cd 0.005, Zn 5.0, Mn 0.1, mg L $^{-1)}$ for drinking water. This situation has arisen as a result of the rapid expansion of industrial activities, followed by increased urbanization and growth of population with exploitation of natural resources, extension of modern agricultural practices as well as the lack of environmental regulations [14]. The present finding showed significant co-relationships between heavy metal concentrations of water and that of the fish tissue, which may indicate that the fish bio-accumulated these elements from the water of Uppanar River. Further, these present result showed that metal concentrations in muscle tissue were below the allowable concentration and consumption safety tolerance in fish set by countries elsewhere, and suggested by WHO. The maximum levels permitted for fishes are - Zn 100, Cu 30, Mn 1.0, and Pb 2.0, µg g $^{-1}$ as per WHO and permitted level Cd is 0.05 – 0.1 µg g $^{-1}$ [14], None of the metals in the present results were above the prescribed limits; thus have, little threat to public health. The present study indicates that, consumption of these species is safe. However, it is quite evident that there was bioaccumulation of heavy metals in fish tissues and condition may get worse. Therefore, a regular monitoring of heavy metal levels in fishes is necessary.

HEAVY METAL CONCENTRATION OF INDIAN RIVERS

The comparison of dissolved metal concentration of River Uppanar with other Indian rivers showed that the Cd, Cu and Pb concentration was several times higher than the Achankoil, Ganga, Brahmani and Mahanadi River (Table 2). The comparative results showed Uppanar River is highly contaminated with industrial effluents discharge, which are the important point sources of toxic heavy metals like Cd, Cu and Pb. The concentration of Zn and Mn at Uppanar River was higher when compared to Mahanadi, Ganga and Brahmani, while it was several hundred times higher than Achankoil River [15, 16, 17, 18 and 19]. The present study revealed that pollutants found in river water are also present in pharmaceutical industrial effluents at higher frequencies of occurrence and concentration. Therefore, pharmaceutical industrial ETPs are clearly a significant point source for organic pollutants in surface waters.

Table 2: Comparison of dissolved metal concentration with other Indian River (µg/L)

Rivers	Cd	Cu	Pb	Zn	Mn	References
Mahanadi River	-	5.9	2.68	11.0	96.9	Kanhauser et al., (1997)
Achankovil River	6.0	224	72	415	699	Prasad et al., (2006)
Ganga River	5	10	120	60	260	Aktar et al., (2010)
Damodhar River	300	3950	-	-	-	Chatterjee et al., (2010)
Brahmani River	4.0	4.7	27	80.1	102	Reza et al., (2010)
Uppanar River	36.08	191.5	98.5	201.38	273.93	Present Study

CONCLUSIONS

As heavy metals are not decomposed biologically, level of these metals, beyond recommended limit, may exist in the river for quite a long distance and it may lead to the long term health-related problems to the people and communities using the water, particularly as a domestic supply source. The result of this work revealed that the mean level of heavy metals discharged into the river has exceeded the maximum permissible limit set by Indian Standards for Drinking Water. The resulting effect is the increase in background level of all the pollutants along the river. It is also observed that level of pollutants in the river during summer season is relatively higher when compared to the other seasons. When the quality of the river is compared with the Indian Standards recommended limits for source of water supply, the river was found to contain some heavy metals above the recommended limits, indicating pollution. Land-use changes in recent years have resulted in a significant deterioration of the water quality of the Uppanar River. Considering the fact that this area is highly populated with many industries and the final drainage of this river ends into the sea, the water quality and pollution status of this river system is of great concern. Therefore, in this study water samples that were taken during four consecutive seasons confirmed that the river has seriously been polluted with Cu,Pb, Mn,Zn and Cd. The result demonstrated that trace elements have originated from various pollutant sources; however, the main anthropogenic sources were industrial wastes, municipal wastes and run-off from agricultural fields.. Heavy metal concentration found in the edible part of fish species are within the WHO permissible limits for human consumption. Thus, there appears to be no immediate threat to the fisheries of the river due to heavy metal contamination. Though, the results indicate that the heavy metal contamination of the river affects the aquatic life including the fish, a scientific method of detoxification of the river water is essential to improve the health of these fish and in turn, the human beings consuming the fishes of the river. Despite mounting urban sprawl of Cuddalore old town in the past decade increased industrialization consequently followed by releasing of untreated industrial effluents into the river played a significant role in polluting the Uppanar River.

REFERENCES

1. Bakare AA, Lateef A, Amuda OS, Afolabi RO. The Aquatic toxicity and characterization of chemical and microbiological constituents of water samples from Oba River, Odo-oba, Nigeria. Asian Journal of Microbiology, Biotechnology and Environmental Sciences 2003; 5, 11–17.

2. Odiete WO. Environmental physiology of animals and pollution 1999; 261. Lagos: Diversified Resources Ltd.

3. Okafor N. Aquatic and waste Microbiology 1985; 1st edn. pp. 1–14. Enugu: Fourth Dimension Publishing Company Ltd.

4. Sudhira HS, Kumar VS. Monitoring of lake water quality in Mysore City. In: International Symposium on Restoration of Lakes and Wetlands: Proceedings of Lake 2000.

5. Adeyemo OK. Consequences of pollution and degradation of Nigerian aquatic environment on fisheries resources 2003; The Environmentalist, 23(4), 297-306.

6. APHA. 1992. American Public Health Association. Standard methods for the examination of water and wastewater. 18th Ed. Washington, D.C.

7.   Hutchinson T H. Reproductive and developmental effects of endocrine disruptors in invertebrates: in vitro and in vivo approaches. Toxicology Letters 2002; 131: 75–81.

8.   Romo-Kroger C M, Kiley JR, Dinator MI, Llona F. Heavy metals in the atmosphere coming from a copper smelter in Chile, Atmospheric Environment 1994; 28, 705–711.

9.   Wu YF, Liu CQ, Tu CL. Atmospheric deposition of metals in TSP of guiyang, PR China. Bulletin of Environmental Contamination and Toxicology 2008; 80 (5), 465-468.

10.   Venugopal T, Giridharan L, Jayaprakash M. Characterization and risk assessment studies of bed sediments of River Adyar-An application of speciation study. International Journal of Environmental Research 2009a; 3 (4), 581-598.

11.   Venugopal T, Giridharan L, Jayaprakash M, Velmurugan PM. A comprehensive geochemical evaluation of the water quality of River Adyar, India. Bulletin of Environmental Contamination and Toxicology 2009b; 82 (2), 211-217.

12.   IPCS Principles and methods for the assessment of risk from essential trace elements 2002. Geneva, World Health Organization, International Programme on Chemical Safety (Environmental Health Criteria 228).

13.   Agency for Toxic substances and Diseases Registry (2000). Toxic Profile for Chromium. Geneva, World Health Organization, International Programme on Chemical Safety (Environmental Health Criteria 228).

14.   Joint FAO/WHO Expert Committee on Food Additives. Evaluation of certain food additives and contaminants. Cambridge, Cambridge University Press, 1982 (WHO Food Additives Series, No. 17).

15.   Konhauser KO, Powell MA, Fyfe WS, Longstaffe F J, Tripathy S. Trace element chemistry of major rivers in Orissa State, India. Environmental Geology 1997; 29 (1- 2), 132-141.

16.   Aktar G, Williams C, David D. The accumulation in soil of cadmium residues from phosphate fertilizers and their effect on the cadmium content of plants. Soil Science 2010; 121, 86–93.

17.   Chatterjee SK, Bhattacharjee I, Chandra G. Water quality assessment near an industrial site of Damodar River, India.

Environmental monitoring and Assessment 2010; 161 (1-4), 177-189.

18. Prasad AL, Iverson A, Liaw. Newer Classification and Regression Tree Techniques: Bagging and Random Forests for Ecological Prediction. Ecosystems 2006; 9:181-199.

19. Reza R, Singh G, Heavy metal contamination and its indexing approach for river water. International Journal of Environment Science and Technology 2010; Vol. 7, No. 4, 2010, pp. 785-792

Coloured Materials in Surface Water in the Sub Arctic Zone: An Overview of its Formation, Properties and Environmental Changes

Egil T. Gjessing

Harestua, Norway

ABSTRACT

Natural organic matter (NOM) is present in most all surface water. This material is governing all chemical and all biological processes in the aquatic environment, and play a practical role in the drinking water industry From an increasing number of international reports, it is clear that the amount of this colored matter is increasing in areas of the northern hemisphere. We ask why and we suggest a combination of the following four reasons: 1) Climate (temperature, humidity, nature and frequency of precipitation); 2) Quality and quantity of precipitation; 3) Nature of catchment (topography and geology), and due to changes in local climate and 4) Quality and intensity of global radiation. In the early 1960s, there were reports from Scandinavia about the decline of coloured matter in lakes. The present increase in colour in our lakes and rivers is partly due to the fact that there are less mineral acids in precipitation. However, change in climate, most probably, plays an even more important role in many regions. As a consequence of the temperature increase, there will also be a change in the amount of precipitation and change in its regional and local distribution. As NOM is "produced" in soil and as the development is based on chemical and microbiological decomposition of plant residues, an increased temperature and more rain will extend the "production-area". The global dimming will also have a significant impact on an increased colour in surface water, as less photo-degradation and less bio-available organic matter is resulting. The positive correlation between the colour increase in surface water and the amount of precipitation, may indicate, that there might be a limited amount of water-extractable coloured material in the catchment. It is argued that that the "production" of the coloured matter will increase and that natural losses, such as "bleaching" etc. will be reduced down flow. Most probably a number of different environmental "mechanisms" are acting simultaneously and/or separately and differently.

INTRODUCTION

Natural organic matter (NOM) is present in all surface water in the Sub-Arctic region and appears as brownishyellow colour. This organic material is governing all chemical and all biological processes in the aquatic environment. Beside the ecological importance of this coloured material, it also plays a practical role in Water Works, when this type of water is used as drinking water.

There are several practical and hygienic reasons why this coloured organic material is not wanted in our drinking water.

From an increasing number of international reports, it is clear that the amount of this coloured matter is increasing in areas of the northern hemisphere.

The purpose of the present article is to discuss the different environmental "elements" that are involved in this colour-increase phenomenon. This will be of importance for the potential need for upgrading of the existing water works and is important for future planning in the drinking water industry.

THE DEVELOPMENT AND NATURE OF NOM/HUMUS/COLOUR IN WATER

As it appear from the sketch below, the natural "syntheses" of NOM (humic substances) is based on plant and plant residues in the soil, where carbohydrates, proteins, lignin and tannins are involved.

These basic groups of organic matter are disintegrated by chemical oxidation and by microorganisms, using the carbon as the main energy source. The chemical and microbiological activity in the soil is indeed complex. All the reactions are dependant on simple "elements" such as: humidity, content of micro nutrients, pH and indeed on temperature. All these different "participants", will play a role in the development of the" final product".

With regard to the time aspects, relative to the humification processes, there is apparently not an agreement among the experts. However, this may depend, as indicated above, on climatic element, such as quality and amount of precipitation, and temperature.

According to this illustration, (see Figure 1) the degradation products and metabolic bi-products result in a "poly-hetero-condensation.

According to Kostychev [2], the decomposition and mineralization of organic matter in soil have their maximum at 35°C and about 35% moisture.

Excessive soil moisture apparently promotes the formation of small molecular-size humus because it prevent the removal of condensation by-product (water) and thereby hinders the "growth" of the molecule.

It appears from Figure 2 and Table 1, that the areas with a potential high content of natural organic colour, represent a significant part of our globe. Table 1 suggests that the amount of organic colour is nearly twice as high here as the World's meaning value.

Several studies on natural organic matter in water (NOM), performed during the last several decades, inspires Petter Wang (2002) to the illustration shown on Figure 3.

Essentially all micro-nutrition's elements are thus associated with the NOM and may be available for organisms in water (Figure 4). Toxic elements such as mercury, copper, lead and cadmium are coupled with NOM in such a way that their toxicity may be reduced.

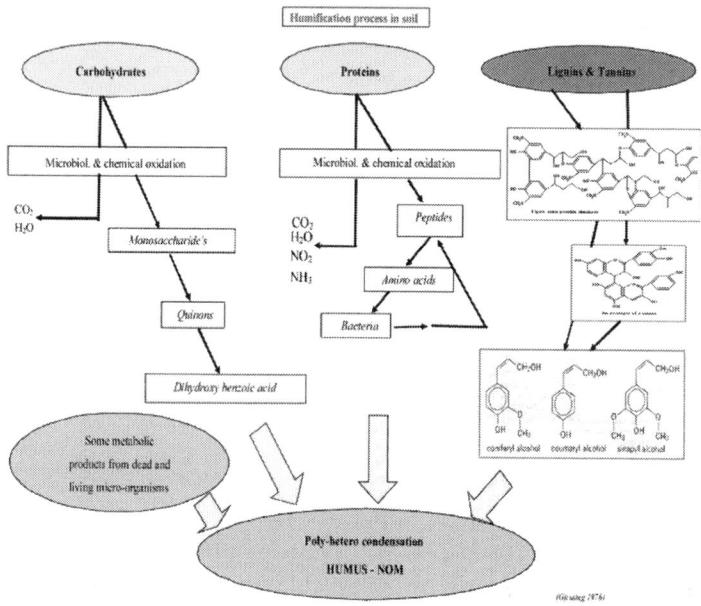

Figure 1: The Humification process [1].

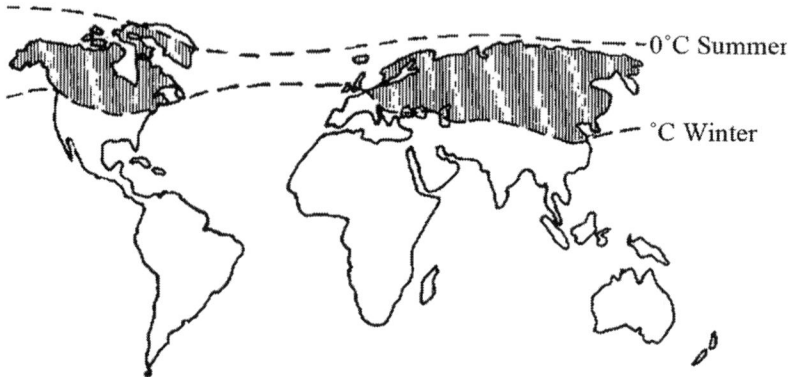

Figure 2: The Sub-Arctic Zone [3].

Table 1: DNOM in surface water in different parts of the world

Climatic zone	DNOM (mg/L)
Temperate	10
Tropical	12
Sub-Artic	20
World mean	11.4

Figure 3: "The Humus Animal" (Petter Wang 2002)[1].

Figure 4: Micro-nutrition's elements are a need for all organisms in water (Petter Wang 2002)[1].

Increased dystrophy reduces the photosynthesis and the photolytic processes in water. This is illustrated on Figure 5.

All processes in nature will be governed by at least the following five "factors": temperature, water, pH, access to micronutrient and light Accordingly, the prognosis will be, that the content and the quality of NOM in surface water will change with the change of climate, quality of precipitation (pH and nutrient, such as N and black carbon [BC]). Global radiation (quality and quantity) will also play a role.

Many years ago, in the early 1960s, there were several reports from Sweden about an increasing number of "clear water lakes". Namely a decrease of colour! At that time, Professor Svante Oden and several colleagues, also from Sweden, were able to show an increase of the acidity of precipitation the Nordic Countries.

During the following years, environmental scientists in Norway and Sweden suggested that the decrease of the colour of the surface water (increasing number of "clear water lakes") could be explained by acid rain. (This was later confirmed by corresponding experts in Norway and elsewhere in the world).

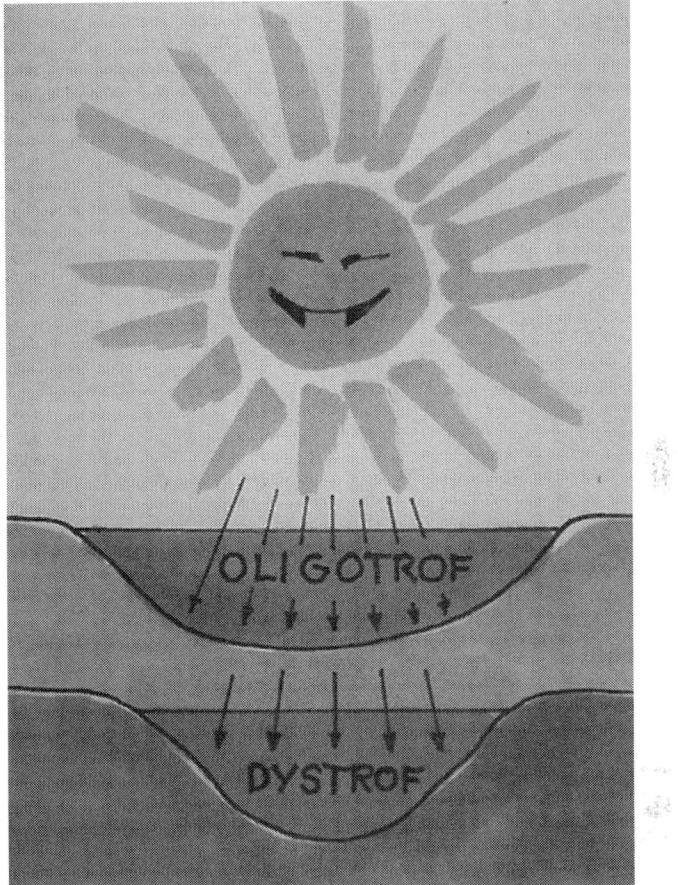

Figure 5: The consequence of a colour-increase is less light and reduced photosynthesis in water (Petter Wang 2002)[1].

The four major reasons for why acid rain reduces the colour in surface water are:

- Reduces microbiological activity in the soil where the humification process takes place (Figure 1).
- Increased adsorption of the coloured organic matter to mineral surfaces in soils and in the water course (see Table 1 and Figure 6).
- Colour decreases with decreasing pH (Figure 7).
- Adsorption of the matter onto Al flocks (Figure 8).

NATURAL QUALITY CHANGES OF NATURAL ORGANIC COLOUR IN WATER

When the water extractable NOM/Colour in the soil, where it is "synthesised", flows into the aquasphere, it will be exposed to another environment The matter will be exposed to: light, air mineral surfaces and a new set of microorganisms. This coloured natural organic matter will get contact with vegetation, in the brook bed, rivers and lakes. All these new environmental "elements" will have an impact on the nature of the NOM.

Adsorption of NOM to minerals (bentonite)

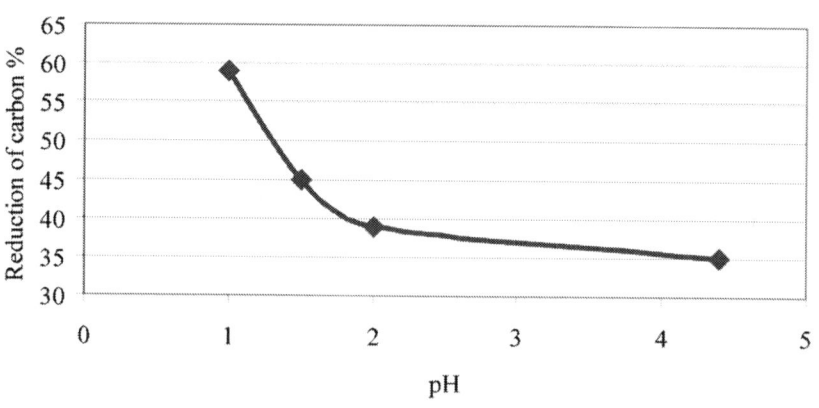

Figure 6: Adsorption of DNOM to Wyoming Bentonite at different pH [1].

Colour related to pH

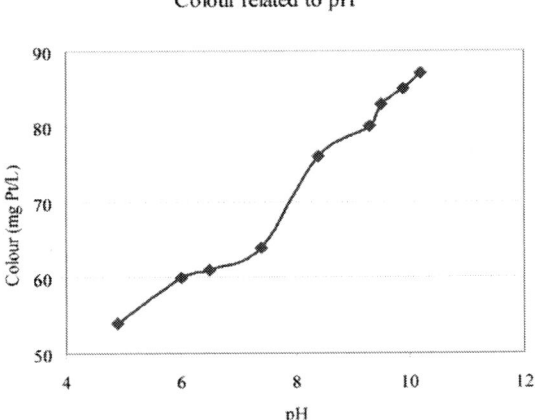

Figure 7: Change of colour with pH [HCl and NaOH was used] [11].

Some studies performed many years ago, illustrate the natural changes of colour in a water course.

The water course of Trehorningen (Baerum near Oslo, Norway) consist of 4 lakes as illustrated on Figure 9 [4].

Several decades ago, samples were taken from the outlets and inlets of these lakes.

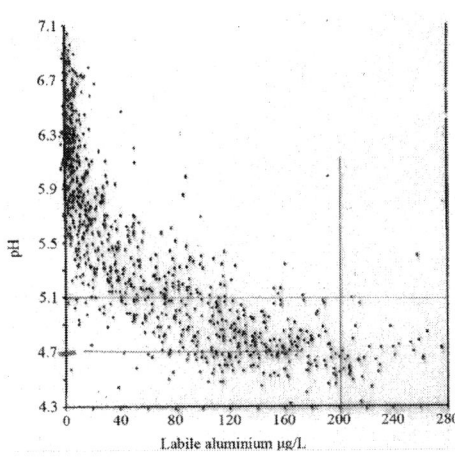

Figure 8: pH and labile Al are strongly correlated in 1005 lakes in Norway [16].

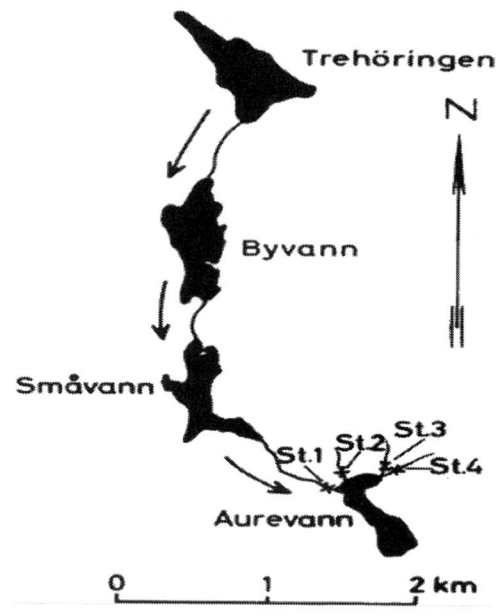

Figure 9: The water course of Trehorningen (Baerum, Norway) [4].

It appears from Figure 10, that there is a colour decrease down-stream; in the lakes and in the rivers, inbetween the lakes (NB! Lake Aurevann is a dammed-up drinking water source, therefore, the increasing colour!).

The colour decrease in the three upper lakes in this study, correlates well with the theoretical retention time for the water This appears from Figure 11. According to these studies the colour reduction in lakes is in the range of 4 colour units per month of storage [4].

Global radiation, in particular the short wave radiation, does have a bleaching effect on the colour in surface water.

Laboratory experiments, with water from the Trehorningen region, and available data on local radiation, suggest that about 20% of this observed colour reduction is due to global radiation [5].

The impact of global radiation on quality of surface water is most probably under-estimated, particularly here in the northern hemisphere. It is well known that short waved radiation of NOM, makes the organic matter more accessible to microorganisms. Several recent studies show that short wave radiation of DNCOM (Dissolved

Natural Coloured Organic Matter) may result in significant changes in important properties of this matter in waterworks [6] The indirect effect of radiation on the "removal" of colour in the lakes should here also be emphasised.

With regard to colour reduction in the streams between the lakes (Figure 10), detailed studies in smaller water systems in the same location, suggest that the mechanisms may be different from than that of the lake.

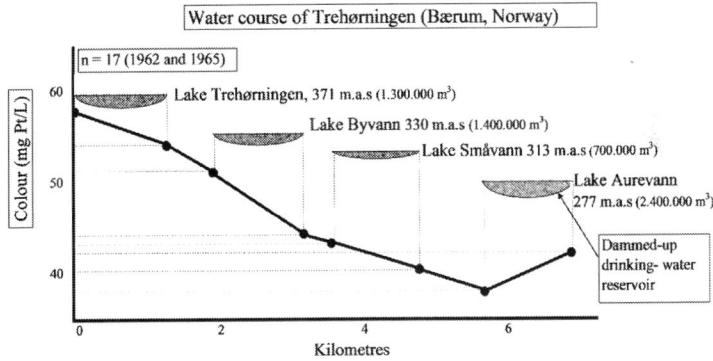

Figure 10: Colour in water downstream the water course of Trehorningen (Baerum, Norway). Samples are taken from outlets and inlets of these lakes through 1962 and 1965 [4].

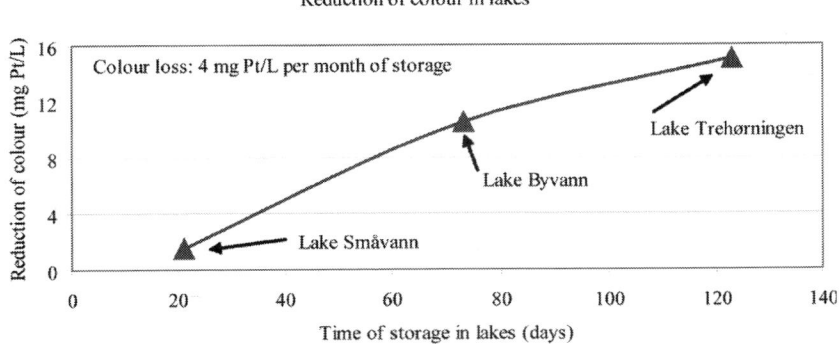

Figure 11: Reduction of colour in lakes (see Figure 9). Colour in the inlets and outlets was measured through 1962 and through 1965. (Number of observa-

tions: [Smavann and Byvann: n = 17; Trehorningen; n = 4]). The colour loss is related to theoretical retention time in the lakes [4].

Samples were taken from a brook, originating from a tarn (Hellerudmyre)[2], through two periods in the early 1960s.

Simultaneously samples were collected 800 metres further down in the same brook (elevation differences about 120 m). The results given in Table 2, show the loss of colour and organic matter and change of quality on this short flow distance, Results from some laboratory experiments, where water from this brook (upper sampling station) is re-circulated in a 30-angel elevated, 2-m-long gutter, which was filled with either crushed stone from this location or crushed inert glass, may suggest that the rate of the water flow has in impact on the result: Higher flow, shorter contact time in the brook and less colour loss [7].

Even though these are simple, laboratory experiments, it could be pointed out that short wave UV-radiation partly oxidise the colour and partly makes it more accessible to adsorption to the mineral surfaces. The short wave UV-radiation (Hg lamp) is believed also to depress the micro-biological activity in the gutter!

Molecular size experiments (both related to the brook studies and to the laboratory experiments) suggest that the high molecular weight fractions, in particular, are lost (Brook studies: MW > 5000; reduced with 67%. MW < 5000 reduced with 38%).

Laboratory experiments: MW > 5000, reduced with 38%. MW < 5000, reduced with 12% [8].

PROPERTIES OF NATURAL ORGANIC COLOUR IN WATER

Recalcitrance

In general the natural coloured matter in water (humic substance, NOM) is considered to be rather stable. However, as illustrated above, physical, chemical and micro biological changes are natural. In general, therefore, it should be emphasised that the BOD (biological oxygen demand) is low.

Solubility

The matter is dissolved water as complex macromolecules. The water solubility will to some extent differ with content of dissolved salt in the water (conductivity), but more important is the role that pH play on the solubility and indeed the properties in general. (Please note the classical definition of Humic Acid, Fulvic Acid etc) [9].

Charge

These organic macromolecules have under natural conditions a net negative charge; the coloured matter act as anione [Humus anion: 3.5 meq/mg DOC]. However, the negative charge will differ with pH.

It will appear from Table 3 that as pH decreases the negative charge of the colour macromolecule will also decrease.

Adsorbability

Under natural conditions the matter will only to a small extent adsorb to physical surfaces. However, again, pH plays a role here. Figure 12 illustrates an experiment where the adsorption of natural colour to mineral (Bentonite) as a function of pH, is studied.

Absorbability

This dissolved natural brownish yellow coloured organic matter absorb light. In water from the same location there will in general be a very good correlation between colour (absorption measured at 420 nm) and DOC and an even better relationship between colour and UV absorption (A_{420nm}). Spectrophotometer measurements are, therefore, widely used as a simple mean to estimate the content of organic matter in water. Please note, however, that that the colour as such do change with pH. This appears from the results given at **Figure 7** [11].

Photosensitivity

Sunlight has in impact on the nature and the amount of coloured material in surface water (see above). The wavelength of the global irradiation in concern in the present publication is wavelength above UV-A (UV-A > 320 nm) It should be mentioned at this point that the intensity of this global irradiation and to some extent the quality do change with altitude.

Table 2: Mean values of a brook in Baerum (near Oslo, Norway): 18 samples taken in 1962 and 1963

	Flow L/sec.	Colour mg Pt/L	DOCb mg C/L	Iron g Fe/L	pH	Colour/ DOC	Colour/ Fe
Brook from Hellerudmyra outlet tam'	1.0	132	12.4	334	4.7	10.6	0.40
800 m down flow (eleva. dif. 120 m)	2.2	51	5.6	149	5.1	9.1	0.34
% loss		61	55	55			

[a]This is the location on which IHSSs Nordic Humic and Fulvic Acid are based; [b]DOC = COD/2.5 - 0.75.

[2]This is the location on which IHSSs Nordic Humic and Fulvic Acid are based (International Humic Substances Society).

Table 3: Per cent colour organic matter that will be electrical neutral in the given pH-range* [10]

pH-range			
1.25-1.50	1.50-1.75	1.75-2.00	2.00-12
18%	52%	12%	18%

*Isoelectric Focusing. Related to Table 3: The NOM (Humus) have am-phyprotic properties. The Isoelectric Focuing Technique (IEF), separate the NOM according to isoelectric point. The NOM sample is mixed with dif-ferent amino acids and different organic and inorganic acids. This is filed into an "Iso-electrical focusing column"— When voltage is applied to the electrodes on both ends of a coloumn containing the mixture, a vertical pH gradient is developed and the NOM migrate until they reach a pH-zone where they are uncharged [10].

Reduction of colour in artificial streams

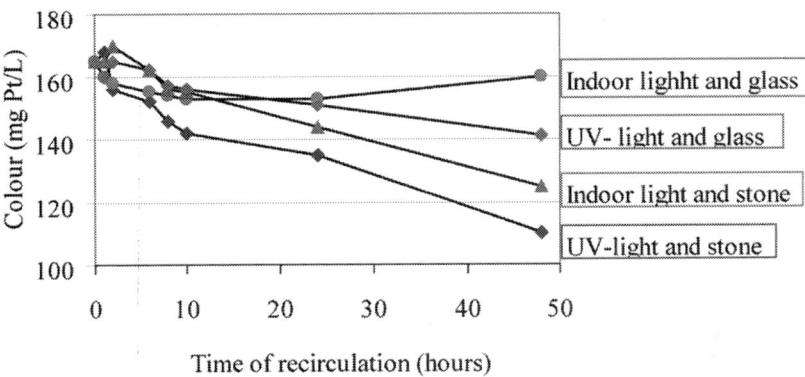

Figure 12: Water from Brook Hellerudmyra (see Table 2). Water is re-circulated in a 30° angel-elevated, 2 m long gutter. This gutter was filled with either crushed stone from the location or crushed glass. The experiment was performed with and without UV-radiation (Hg-lamp) [7].

Even though the present topic is related to the natural environment, I include here some aspects related to natural coloured material and UV-C, because the concern related to the present environmental changes is of importance for practical problems affecting the drinking water industry. When Rook in the early 1970s [12] reported that chlorination of coloured water developed chloroform in drinking water, UV-C disinfection became an important alternative.

However, even though UV-C is an effective method to kill bacteria, viruses and parasites, the treatment does result in a change in bioavailability of the coloured matter in the water. This change in the quality of the DOM showed to give more bioactive material in the water. Depending on the "UV-dose", the resulting "product" acts as an algaecide [13] or a bactericide [14].

Recently it has been demonstrated that, after UV-irradiation of NOM, specific bacterial members are well adapted to low pH, high concentration of low molecular weight DOC and oxidative stress and, therefore, thrive well after the UV-treatment [15]. This may have consequences for the drinking water quality during transport from the treatment plant to the consumer.

Molecular Weight Distribution

The size of these complex macromolecules does indeed change with time of storage in lakes and rivers and with a number of environmental "factors", such as pH [11].

ENVIRONMENTAL DEVELOPMENT IN THE 1970S

An increased acidity of precipitation was first reported in 1958, by Evil Gorham. It was "re-discovered" in the early 1960s by Svante Oden from Sweden.

During a court hearing in Norway in 1965, a hearing related to a dramatic decline in the inland fish-catch in a lucrative Norwegian salmon river, acid rain was mentioned for the third time. Based on some intensive river quality surveys, in the southernmost part Norway, showing significant acidification of major river in this part of Norway, an inter-dispensary research project was established in 1972.

Soil, forest, water and fish were the major "objects" in these studies (the SNSF-Project).

The results of some studies of soil water chemistry, given in Table 4, demonstrate the soil-acidification problem; more than a tripling of H^+-concentration.

The results from this, eventually international research project, revealed, that, besides the direct bio-toxic effect of H^+ in water, the acid rain, resulted in an increase in dissolved Al in water had also indeed a significant negative bio-effect.

The relation between Al and pH is illustrated on Figure 8.

Finally, in relation to the present topic, the development of natural organic colour in soil (and water) it is relevant to emphasise the importance of micro-organisms.

The results on Figure 13 illustrate the impact of pH on the disintegration of leaves.

During the following decades, the monitoring of water quality worldwide and the knowledge about the quality and fate of harmful environmental elements, have indeed increased.

The cleaning of the combustion product, world-wide, has resulted in less polluted precipitation relative to particularly H_2SO_4 (Table 5).

Table 4: Change of acidity in soil in Rondane (Norway) between the 1940s* and 1984

Depth	Mean pH		Mean H+	
	1945	1984	1945*	1984
A0	4.37	4.04	43	91
A2	4.92	4.34	12	46
Bh	5.14	4.61	7	25
Bf	5.49	4.75	3	18
C	5.69	4.84	2	14

*1942-1949

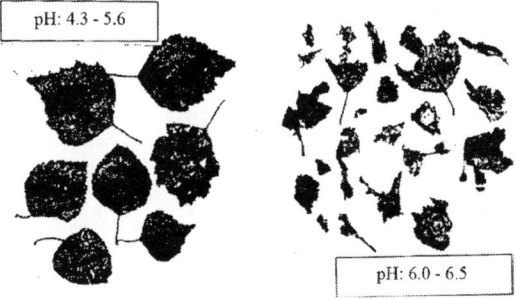

Figure 13 :Birch leaves from litterbags that have been left for one year in acid waters (pH 4.3 - 5.6) [weight loss: 46%] compared to leaves in more neutral waters (pH 6.0 - 6.5) [weight loss: 52%] in southern Norway [17].

Table 5: Change in quality of surface water in Norway between the mid 1980s and the mid 1990s. Unit: μeq/L

Norway	Number	SO4			H+		Al (inorg.)	
	of lakes	1986	1995	1986	1995	1986	1995
East	65	73	50	3.6	2.2	42	28
South	189	60	36	15.1	10.7	112	64
West	94	30	23	5.8	4.1	22	15
Mid	59	16	12	5.8	0.7	4	1
North	78	47	42	0.3	0.3	0	0

A remarkable consequence of this reduction in H_2SO_4, is an increase in the salmon catch in many rivers in the effected regions of Norway.

The monitoring of the surface water quality has also revealed an increase in the colour in many areas in the northern hemisphere. This finding was first reported by Forsberg and Petersen in Sweden in the late 1980s [18].

During the last decades there have been an increased number of reports on an increasing colour in surface water in the sub-Arctic Zone. An example is illustrated in Figure 14.

In general the phenomena of an increase of colour in surface water, apparently started most places, in the mid 1970s.

WHY DOES THE COLOUR INCREASE?

The impact of climate change on Dissolved Natural Coloured Organic Matter (DNCOM) is believed to mainly be related to temperature, to the amount and quality of precipitation and to global irradiation.

Temperature Increase

- Increase in general growth;
- Change in quality of vegetation;
- More litter;
- Higher rate of decomposition;
- Increase in biochemical activity.

Colour [Farris. South East Norway]

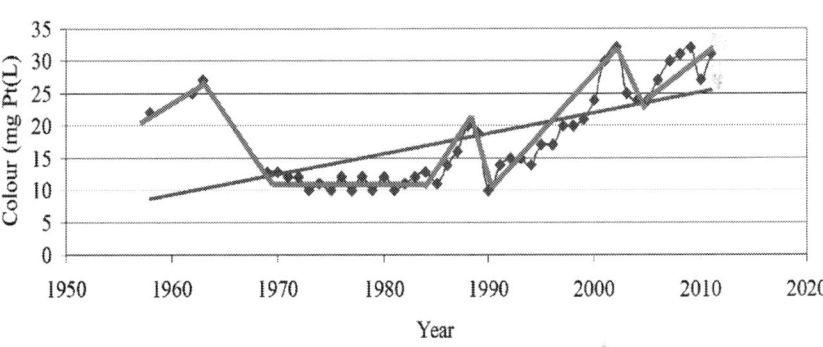

Figure 14: Colour in Lake Farris during the last 50 years.

During the last several decades, the tree-line has visibly retreated up-hill and so has also the nature of the vegetation. It appears from Figure 1, illustrating the biochemical processes in soil, that a change in the "input" may have an impact on the quality of the final product. Likewise an increased temperature will result in a higher rate of decomposition of dead plant material and indeed increase and change the quality of the biochemical activity.

Change of Amount of Precipitation

- Expansion of peat-land area;
- Increased humidity in the soil (favouring smaller molecules);
- More (DNCOM) extracted from soil into water;
- Shorter time of storage of water down-flow.

Reports from the international community show an increase in both the amount and the regional distribution of precipitation.

An increase in precipitation will expand the peaty-land area and consequently increase both the production of and the water extraction of DNOCM.

Another important consequence of more water is shorter retention in the catchment area.

A shorter time of storage in both the terrestrial and aquatic phase, will also, as demonstrated above, have an impact on the nature and the amount of DNOCM. There will be a shorter time for chemical, biochemical and photochemical reactions.

Change of Quality of Precipitation

- Increased pH;
- Increase in colour (see Figure 7);
- Increased microbiological activity (in soil and water) {see Figure 13};
- Reduced S/N-ratio;
- Less sulphur (S) in DNCOM;
- Increase in BC (Black Carbon).

The impact of the concentration of H^+ on all processes in life is remarkable. It is also remarkable to notice that it is only a few decades ago, since we realized that the quality of precipitation had to be accounted for relative to the quality of water. Here we will limit our discussion to pH in rain and snow, except for one relevant observation regarding cations in rain and snow and exchange of H^+.

It appears from Figure 15 that cations in precipitation will decrease pH in soil and water.

Regarding polluted precipitation and quality of DNCOM, it is first of all the mineral acids H_2SO_4 and HNO_3 that will be included here.

The direct effect of increased pH on colour (Figure 7), on molecular size (Figure 16) and on charge (Table 3) Related to Figure 16, the samples are fractionated on. Sephadex G-25 gel. Each point on the figure, is the mean value of 4 - 12 different fractionations. The following chemicals were used to adjust the salt content and pH: NaCl, Na_2SO_4,

HCl, NH_3, H_2SO_4, NaOH. will, as stated before, point in the direction of increased colour in water.

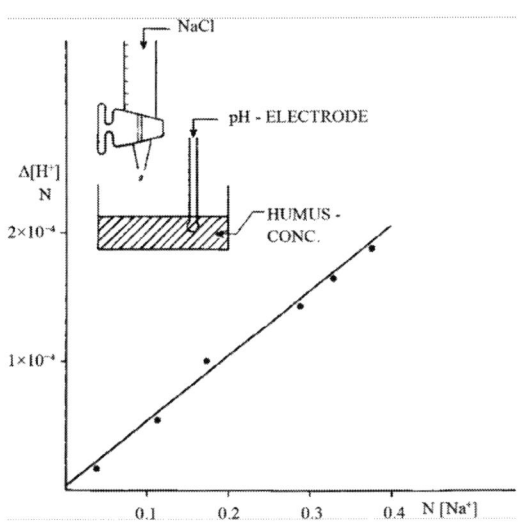

Figure 15: DNCOM acts as a cation exchanger [19].

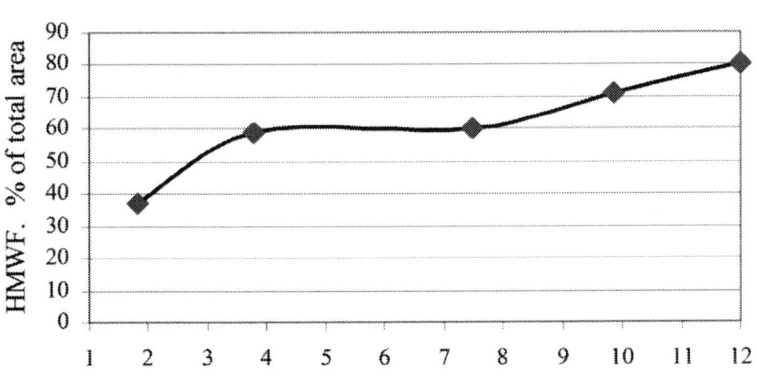

Figure 16: Effect of pH on the molecular size of coloured organic matter in water [11].

The reduction of mineral acids in precipitation, particular sulphuric acid, will, in addition to the impact on the micro-biological activity (also in soil) possibly result in a less sulphur "imbedded" in DNCOM. The results from some earlier work, where the content of organic sulphur is studied, suggest an effect of H_2SO_4 on the composition of the DNCOM. This work suggested 1) "An esterification of the coloured matter"; 2) That the S apparently is connected to the HMWF (high molecular weight fraction) and 3) That this sulphur-contained DNCOM apparently acts more toxic to Daphnia and Yearlings of Salmon [20].

In the de-acidification of surface water, H_2SO_4 does play the most dominant role.

Because the energy production in the coming decades will still be based on burning, in an N-contained atmosphere (78% N; 21% O) Consequently the producing of NO_x will continue to increase. (N_2 + O_2 + heat = NO_x).

In addition to the positive effect that this decrease in the S/N-ration in the precipitation may have on the biochemical processes in the catchment, the fertilising properties of BC (Black Carbon) should be mentioned. This black matter will also have an impact on the albedo, regarding both regarding snowmelt during spring and on surface temperature in general [21].

Change in Global Irradiation

Less Light Reaches the Floor

As a consequence of inreased vegetation in the catchment, less light penetrate down to the surface water and reduces the action of the radiation.

Global Dimming

Global dimming is the gradual reduction in the amount of global direct irradiation at the Earth's surface. This was observed for several decades after the start of systematic measurements back in the 1950s [21].

It is thought that global dimming is due the increased presence of aerosol particles in the atmosphere, caused by human action. Aerosols and other particulates absorb solar energy and reflect sunlight back into space. The pollutants can also become nuclei for cloud droplets. Water droplets in clouds coalesce around the particle. Increased pollution causes more particulates and thereby creates clouds consisting of a greater number of small droplets. This means that the same amount of water is spread over more droplets. Small droplets make the clouds more reflective, so that more of the incoming sunlight is reflected back into space and less reaches the Earths surface. The degree of this reduction in irradiation differs in the international reports and indeed it is emphasised that there are local differences; such as influences from large cities [22]. Percent wise a decrease in the range of 2% - 4% per decade is reported. A report from Israel, based on 50 years of operation, conclude with a globally average reduction in irradiation of 0.5 W/m^2 per year, equivalent to a reduction of 2.7% per decade [23].

Relative to colour in water, the importance of photochemical action for the "production" of more accessible food for aquatic organisms should again be emphasised.

Reduced Bleaching and Photo Degradation

During the last decades, there has been an increasing awareness of the role that short wave radiation plays for processes in water As pointed out, the DNCOM is rather recalcitrant. However, when exposed to light, photo chemical reactions result in bleaching and probably most important, making the DNCOM more bio-available. Global dimming will alter these processes!

CONCLUDING REMARK

The ecological role that the Dissolved Natural Coloured Organic Matter (DNCOM), plays in the aquatic environment, is essential.

In the aqua sphere this matter is a result of natural processes in the terrestrial environment, starting with the photosynthesis, involving light, CO_2, chlorophyll and nutrients.

For a number of reasons the content of DNCOM in water will depend on the climate: basically temperature and precipitation (humidity), that do govern all ecological processes.

In the late 1980s Forsberg and Petersen [18] observed an increase in colour in lake water in Sweden.

According to their reports, this increasing colour started in the mid 1970s.

As fare as I know, Forsberg and Petersen showed these results for the first time.

Along with this observed increase in colour there was also a decrease in transparency of the water.

During the following few decades' similar reports were published from other regions in the Sub Arctic Zone, suggesting that this may be a world-wide phenomena.

Expert ecologists have basically used the global climatic changes and change in quality of rain and snow as the explanations.

The ecological importance of this change in water quality relates essentially to an increase in the distribution of micro-pollutants in the environment due the complexing ability of this coloured matter ("polluted DNCOM" (see Figures 3 and 4)) and the "dystrophication" of the aquatic environment (less light penetrates the water (see Figure 17).

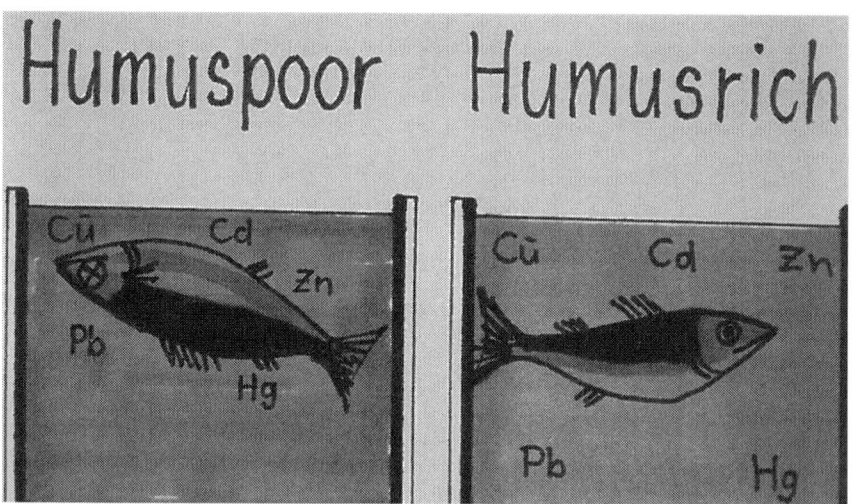

Figure 17: Elements, toxic to aquatic organisms may be complexed to the coloured matter, resulting in reduced action[3].

However, as always, the direct economical consequences are what trigger further efforts.

During the last 11/2 decades, the NOM in drinking water has received more and more hygienic attention. This is partly due to the "polluted nature" of the coloured material and partly due to microbial transformation of the DNCOM in its increasing distance between the source and the consumer. The drinking water industry needs to know about future investments with regard to colour removal.

Climatic changes involve both short-term and the longterm weather conditions.

In relation to quality of surface water and increase in colour, it is basically:

- temperature
- amount and quality of precipitation and
- amount and quality of global radiation that have an impact on quality and amount of the matter.

Based on our analysis of available data, based on the prognosis given by experts within climatic changes and based on own accumulated knowledge, I may conclude that the colour in surface water will continue to increase in many areas of the Northern Hemosphere.

REFERENCES

1. Gjessing, E.T. (1976) Physical and chemical characteristics of aquatic humus. Ann Arbor Science Publishers Inc., Ann Arbor.

2. Kostyckev, P.A. (1886) Soil of the chernozem region of Russia. Part I. The formation of chernozem. In: Aleksandrova, L.N., Ed., Study of the Humification of Plant and the Nature of Newly Founded Humic Acid, Soviet Soil Science, 429-437.

3. Gjessing, E.T. (1980) Water treatment considerations— Aquatic humus. In: Smith, D.W., Ed., Proceeding from Post Conference in Alberta, Canada. International Association on Water Pollution Research, Oxford, 95-101.

4. Gjessing, E.T. and Samdal, J.E. (1968) Humic substances in water and the effect of impoundment. Journal (American Water Works Association), 60, 451-455.

5.   Gjessing, E.T. and Gjerdal, T. (1970) Influence of ultraviolet radiation on aquatic humus. Vatten, 26, 144-145.

6.   Paul, A., Dziallas, C., Zwirnmann, E., Gjessing, E.T. and Grossart, H.P. (2012). UV-irradiated freshwater: Impact of Natural Organic Matter (NOM) on quality of organic matter and on bacteria. Aquatic Sciences, 74, 443-454. doi:10.1007/s00027-011-0239-y

7.   Gjessing, E.T. (1970) Reduction of aquatic humus in streams. Vatten, 26, 14-23.

8.   Gjessing, E.T. (1970) Some factors affecting on the stability of aquatic humus. Vatten, 26, 135-143.

9.   Oden, S. (1910) Die Huminsauren Kolloidchem. Beiheft, 11, 75-98.

10.   Gjessing, E.T. and Gjerdahl, T. (1975) Electromobility of aquatic humus. Fractionation by the use of the isoelectric focusing technique. In: Povoledo, D. and Golterman, H.L., Eds., Humic Substances—Their Structure and Function in the Biosphere. Proceedings of an International Meeting, Nieuwersluis, 29-31 May 1972, 43-51.

11.   Gjessing, E.T. (1971) Effects of pH on the filtration of aquatic humus using gels and membranes. Schweizriche Zeitscherift fur Hydrologie, 33, 592-600.

12.   Rook J.J. (1974) Formation of haloforms during chlorination of natural waters. Water Treatment Examination, 23, 234- 243.

13.   Gjessing, E.T. and Källqvist, T. (1991) Algicidal and chemical effect of u.v.-radiation of water containing humic substances. Water Research, 25, 491-494. doi:10.1016/0043-1354(91)90087-7

14.   Paul, A., Gjessing, E.T., Lønnechen, H. B. and Liltved, H. (2006) Bactericidal effect of water containing natural organic matter (NOM). In: Fritz, H.F. and Abbt-Braun, G., Eds., Humic Substances—Linking Structures to Functions, Proceeding from 13th Meeting of the IHSS, Karlsruhe, 921-924.

15.   Paul, A., Dziallas, C., Zwirnmann, E., Gjessing, E.T. and Grossart, H.P. (2012) UV-irradiated freshwater: Impact of natural organic matter (NOM) on quality of organic matter and on bacteria. Aquatic Sciences, 74, 443-454. doi:10.1007/s00027-011-0239-y

16.   Rosseland, B.O. and Henriksen, A. (1990) Acidification in Norway. Loss of fish population and the 1000-lake survey. Science of The Total Environment, 96, 45-56.

17. Traaen, T. (1980) Personal communication. Norwegian Institute for Water Research, Oslo.

18. Forsberg, C. and Petersen, R.C. (1990) A darkening of Swedish lakes due to increased humus input during the last 15 years. Verhandlungen der Internationalen Verein Limnolgie, 24, 289-292.

19. Gjessing, E.T. and Johannessen, M. (1976) Potential effects of metals in precipitation on the exchangeable humus-hydrogen in soil and surface water. In: Nriagu, J.O., Ed., Environmental Biogeochemistry: Vol. 2. Metal Transfer and Ecological Mass Balances, Ann Arbor Sciences Publishers Inc., Ann Arbor, 557-563.

20. Gjessing, E.T., Efraimsen, H., Grande, M., Källqvist, T. and Riise, G. (1991) Changes in properties of humic substances by sulphuric acids acidification. In: Baker, R.A., Ed., Organic Substances in Sediments and in Water, Lewis Publishers, Michigan, 89-98.

21. Wikipedia. www.Wikipediano/

22. Abakumova, G.M., Feigel, E.M., Russak, V. and Stadnik, V.V. (1996) Evaluation of long-term changes in radiation, cloudiness and surface temperature on the territory of the former Soviet Union. Journal of Climate, 9, 1319-1327. doi:10.1175/1520-0442(1996)009<1319:EOLTCI>2.0.CO;2

23. Stanhill, G. and Cohen, G. (2001) Global dimming: A review of the evidence for a widespread and significant reduction in global reduction with discussion of its probable causes on possible agriculture consequences. Agricultural and Forest Meteorology, 107, 255-278. doi:10.1016/S0168-1923(00)00241-0

Water Pollution with Special Reference to Pesticide Contamination in India

Anju Agrawal[1], Ravi S. Pandey[2],
and Bechan Sharma[3]

[1]Department of Zoology, Surendra Nath Balika Vidyalaya Post Graduate College, CSJM University, Kanpur, India
[2]Department of Zoology, University of Allahabad, Allahabad, India
[3]Department of Biochemistry University of Allahabad, Allahabad, India

ABSTRACT

The pesticides belong to a category of chemicals used worldwide as herbicides, insecticides, fungicides, rodenticides, molluscicides, nematicides, and plant growth regulators in order to control weeds, pests and diseases in crops as well as for health care of humans and animals. The positive aspect of application of pesticides renders enhanced crop/food productivity and drastic reduction of vector-borne diseases. However, their unregulated and indiscriminate applications have raised serious concerns about the entire environment in general and the health of humans, birds and animals in particular. Despite ban on application of some of the environmentally persistent and least biodegradable pesticides (like organochlorines) in many countries, their use is ever on rise. Pesticides cause serious health hazards to living systems because of their rapid fat solubility and bioaccumulation in non-target organisms. Even at low concentration, pesticides may exert several adverse effects, which could be monitored at biochemical, molecular or behavioral levels. The factors affecting water pollution with pesticides and their residues include drainage, rainfall, microbial activity, soil temperature, treatment surface, application rate as well as the solubility, mobility and half-life of pesticides. In India organochlorine insecticides such as DDT and HCH constitute more than 70% of the pesticides used at present. Reports from Delhi, Bhopal and other cities and some rural areas have indicated presence of significant level of pesticides in fresh water systems as well as bottled drinking mineral water samples. The effects of pesticides pollution in riverine systems and drinking water in India has been discussed in this review.

INTRODUCTION

Water is essential for life. No living being on the planet Earth can survive without it. The major part of water on earth is marine water which cannot be used without processing by human beings. The only available fresh water which could be used for drinking purposes arises from the ground water. The percent volume of it, however, is sufficient to cater the need of the living beings, provided it would have been of high quality. Water quality is important in our lives because it is essential to support physiological activities of any biological cell.

Water pollution may be defined as any impairment in its native characteristics by addition of anthropogenic contaminants to the extent that it either cannot serve to humans for drinking purposes and/or to support the biotic communities, such as fish. Water pollution is the contamination of water bodies such as lakes, rivers, oceans, and groundwater by human activities. All water pollution affects organisms and plants that live in these water bodies and in almost all cases the effect is damaging either to individual species and populations but also to the natural biological communities. It occurs when pollutants are discharged directly or indirectly into water bodies without adequate treatment to remove harmful constituents.

Water pollution is a major cause of global concern as it leads to onset of numerous fatal diseases [1] which is responsible for the death of over 14,000 people every day. The problem in developing countries is more alarming than that of industrialized nations. In addition to pesticides, natural phenomena such as volcanoes, algae blooms, storms, and earthquakes also cause major changes in water quality and the ecological status of water. Water pollution has many causes and characteristics. If the quality of water is changed by the presence of toxins, it becomes potentially harmful to these life forms, instead of sustaining them.

Many water pollutants are reported to act as toxic chemicals. The pesticides are designed and developed keeping in view killing the insects-pests in general and they are not specific-specific. Their application methodologies are designed to ensure that these chemicals come in contact with the target pests to kill them avoiding the non-target organisms. These target pests, however, are simply species of animals that share many of the same characteristics of other animals. One of these characteristics is a susceptibility to certain toxins. In other words, a chemical that is toxic to one animal also may be toxic to other forms of animal life. Although it might take a larger dose of pesticide to harm humans than pests such as insects, many pesticides are still toxic to humans. The doses needed to kill a pest effects the humans in many ways such as disruption in function of sex hormones and reproductory performance [2-5]. The pesticides act as xenohormones (mimicking the action of endogenous hormones) or otherwise interfering with endocrine processes, hence have been collectively categorised as endocrine disruptors [6].

An herbicide is a substance used to kill unwanted plants. Selective herbicides kill specific targets while leaving the desired crop relatively unharmed. Some of these act by interfering with the growth of the weed and are often synthetic "imitations" of plant hormones. Herbicides used to clear waste ground, industrial sites, railways and railway embankments are non-selective and kill all plant material with which they come into contact. Smaller quantities are used in forestry, pasture systems, and management of areas set aside as wildlife habitat. Many of them are species specific to the target plant pests [5].The exceptions to this are broad-spectrum herbicides that are designed to kill a wide variety of plants. An herbicide that is specific to one or more species of plants does not ensure that it is safe to enter the water system. Some of the dangers from these chemicals are yet to be fully understood. Caution should therefore be used to ensure that these products do not unnecessarily enter the water system. Using safe, well-planned applications of materials, such as pesticides, the risk to humans and other animals is minimal. If these products enter the water system, they may reach nontarget animals and pose a hazard to the lives of other animals (including humans and domestic animals) and non-target plants. Along with pesticides, there are many other materials that can cause the same type of adverse effects to water systems and ultimately to humans [5]. The most reasonable way to deal with the problem of water pollution could be by striving not to introduce any hazardous materials into waters without reason, because the result may be a deterioration of water quality. The overall picture is not as bleak as it appears. As the threat to water systems and the mechanisms that cause water to become polluted are now better understood, steps are needed to protect the quality of our water. Keeping the seriousness of pesticides contamination in water systems and its impact on humans and animals in addition to the environment, an endeavor has been made in the present review to compile and project the current information available on this issue with special reference to India.

SOURCES OF WATER POLLUTION

Water pollution is the contamination of water bodies (e.g. lakes, rivers, oceans, and groundwater). This may be defined in terms of the undesirable changes in the chemical and physical properties of water which are not favourable to all those living things utilizing water for their lives. There are two basic forms of water pollution; 1) changing the types and amounts of materials carried by water, and 2) altering the physical characteristics of a body of water [7]. Water pollution occurs in many forms, from a wide range of sources. Agriculture may contribute to water pollution from feedlots, pastures, and croplands. Mining, petroleum drilling, and landfills may also be major sources of water pollution. Other water pollution sources, related to humans, are sanitary sewers, storm sewers, industry, and construction [5].

According to one report published in 1990 from the Environment Protection Agency (EPA), > 50% of the water pollution of streams and rivers occur due to leaching and mixing of chemicals from the agriculture practices [5]. The next highest source was municipal sources (about 12%). Groundwater contamination is from several sources (USGS Circular 1998), including agricultural activities, storage tank leakage, industrial waste, sewer and septic leakage, leaching from landfills, mining, and many other sources. Water pollution occurs when a body of water is adversely affected due to the addition of large amounts of materials to the water. The sources of water pollution are categorized as being a point source or a non-source point of pollution. Point sources of pollution occur when the polluting substance is emitted directly into the waterway [8]. A pipe spewing toxic chemicals directly into a river is an example. A nonpoint source occurs when there is runoff of pollutants into a waterway, for instance when fertilizer and pesticide from a field is carried into a stream by surface runoff. A toxic substance is a chemical pollutant that is not a naturally occurring substance in aquatic ecosystems. The greatest contributors to toxic pollution are herbicides, pesticides and industrial compounds.

Pesticides are those chemicals (such as insecticides, fungicides, herbicides, rodenticides, molluscicides, nematocides, plant growth etc.), which have been widely used throughout the world to increase crop yield and to kill the insect-pests responsible for transmitting various diseases to humans and animals. However, according to several reports, these chemicals have been proved to inflict adverse impacts on the health of living beings and their environment [9-12].

In most of the technologically advanced countries, organochlorine (OC) insecticides, which were used successfully in controlling a number of diseases such as malaria and typhus, have been banned or restricted. After 1960, other synthetic insecticides such as organophosphate (OP), carbamates, pyrethroids, and herbicides and fungicides were introduced into agricultural practices as well as several health management programmes.

The trend of application of different pesticides in India radically differs from rest of the world. The data presented in Figures 1(a) and (b) reflects the estimates of global usage of pesticides (Figure 1(a)) in general and India (Figure 1(b)) in particular. The 76% of the total pesticides used in India is insecticide (Figure 1(b)). Correspondingly, the lesser use of herbicides and fungicides is reflected (Figure 1(b)). The main use of pesticides in India is for cotton crops (45%), followed by paddy and wheat [11]. The pesticide cycle is illustrated in Figure 2.

The major part of the pesticides applied in any area for a specific reason (about 99%) remain unused and it gets mixed with air, soil, water and plants which by several means causes harmful effects on the people, pets, and the environment. Not only the farmers in rural areas but also the people in urban areas use more than half of pesticide in their homes and home gardeners, in and around the schools, business areas, and hospitals etc.

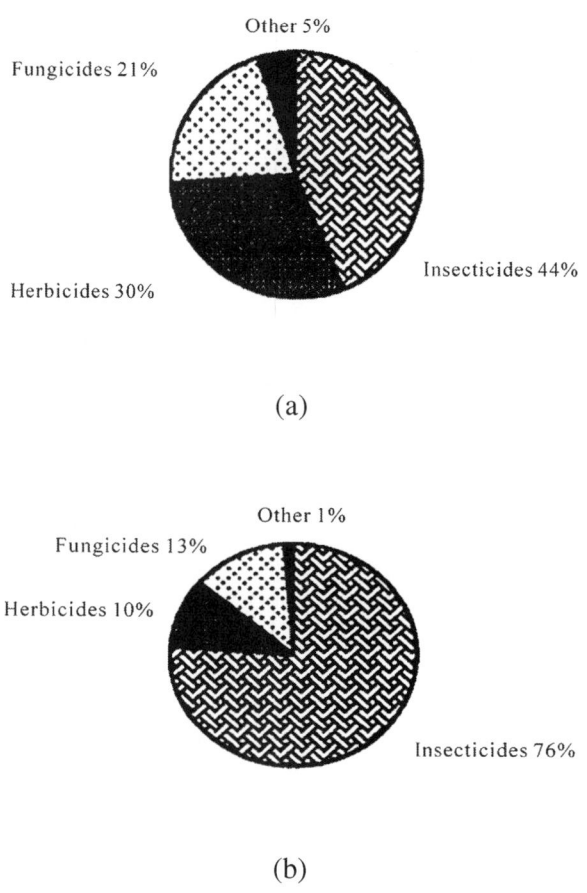

(a)

(b)

Figure 1: (a) An estimate representing application of different pesticides globally [11]; (b) scenario of application of different pesticides in India [11].

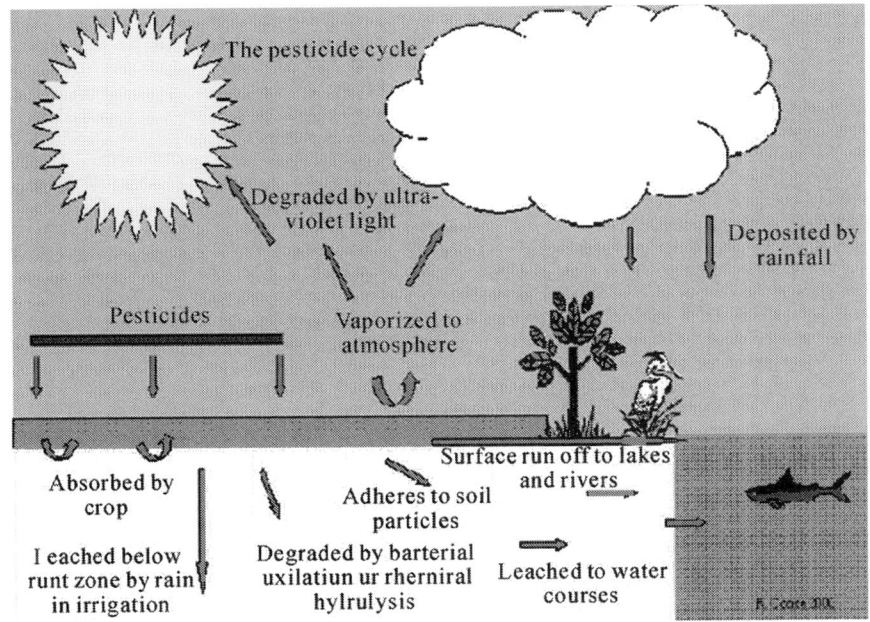

Figure 2: A scheme showing the different stages involved in pesticide cycle, source: website: The University of Reading, ECIFM, pesticides.

A pesticide is that compound which should be lethal to the targeted pests only and not to the non-target living organisms such as humans and animals. But the disproportionate application of these compounds has adversely affected the flora and fauna of the entire ecosystem. After the death of about 100 people in India due to consumption of parathion contaminated wheat flour [12], Indian Council of Agricultural Research (ICAR) constituted a committee to suggest possible remedies to combat the toxicity caused due to presence of pesticides and their residues in the edibles [13]. After the first warning about the poisoning of organochlorines (OC) to living systems [14], the reports from US National Academy of Sciences endorsed the same by studying the toxicity of OC compounds and their metabolites in birds [15]. The pesticides have been shown to display their effects by causing xenotoxicity, alterations in body's immunity, reproductive system and other physiological processes of different organisms thereby generating several diseases including cancer [16-18].

PROPERTIES OF PESTICIDES

The role of pesticides is to kill the insect-pests, but this property of pesticides makes them a poison to other organisms including different birds, fish species, animals and humans. These pesticides are not target specific. The constant exposure of pesticides to non-target species may lead to induce toxicity once it crosses the threshold limit in the system. It is known that the major portion of the pesticide applied in an area reaches into healthy environmental components such as aquatic reserves (ponds, lakes, rivers and oceans), where they gradually get accumulated into other organisms [5].

Potential Chemical Component of the Pesticides

The potentially active ingredient of any pesticide is the chemical compound that brings about the desired effect. In the case of a pesticide, the active ingredient is the material intended to kill the target pest and has the potential to be dangerous to other animals. The other substances in a pesticide are usually inert (not reactive) and are used to carry the toxin (active ingredient) while making its application easier. The active ingredient is usually a very small percentage of the total ingredients in a pesticide [5].

Toxicity Bioassay

The pesticide toxicity is the degree to which it is able to damage an exposed organism. Toxicity can refer to the effect on a whole organism, such as an animal, bacterium, or plant, as well as the effect on a substructure of the organism, such as a cell (cytotoxicity) or an organ (organotoxicity), such as the liver (hepatotoxicity). These terms are defined for toxicity to specified animals after exposure of a specified time. These toxicity terms can apply to target pests or non-target animals, including humans. The most common of these toxicity terms are LD_{50} and LC_{50}.

Lethal Dose Determination

According to Cook et al. [5], the LD_{50} is a measure of a substance's toxicity. LD_{50} stands for the dose of a substance, such as a pesticide, that kills one-half of the animals tested. The LD_{50} for a specified animal is the amount that must be in or on the body of that type of animal to kill half of the affected population within a given amount of time. When the LD_{50} of chemicals in animals is compared it gives a relative ranking of the toxicity to each animal. LD_{50}'s are often calculated using rats, because humans cannot be tested in a way that will test how many are killed, given a certain dose. This information from LD_{50} is calculated for rats and can be used to estimate the LD_{50} for humans by multiplying by 70 (the average kilogram mass of humans). Substances that are toxic to one mammal are often toxic to another. This conversion is an estimate that might not accurately calculate limits for human exposure. The comparative toxicity of pesticides is shown in Table 1.

Lethal Concentration Determination

LC_{50} stands for the lethal concentration of a material to kill one-half of the animals tested in a specified amount of time. It is the amount of a material that comes in contact with the animal being tested that will kill one-half the population affected. This lethal concentration may be in a medium such as the air or a body of water. In this context, it will deal with the amount of a substance in water that would kill animals that live in that body of water. In other words, if the LC_{50} is present for a type of fish, then the concentration of a toxin in the water is at a level that will kill one-half of that type of fish that are present in that body of water. Some commonly used insecticides are given with their properties and LC_{50} for fish in Table 2.

Pesticide Formulation

The main purpose of pesticide formulation is to manufacture a product which has optimum biological effi ciency, is convenient to use, and minimizes environmental impacts. Active ingredients are mixed with solvents, adjuvants (boosters), and fillers as necessary to achieve the desired formulation. Pesticides may be in several physical forms or formulations. They may be water dispersible granules, dusts, aerosols, emulsifiable concentrates, flowable concentrates, solutions, solid baits, or liquid baits. They are sold in these forms because of advantages they offer to their application. Formulations influence the deposition on the soil or plant surface. In turn, they may regulate or influence its uptake by the plant or its movement into the upper soil profile. Formulations also determine the wash off or runoff characteristics of a pesticide in rain or irrigation water [5].

Pesticide Efficacy

Cook et al. [5] reported that the effective dose is the amount needed to kill a target pest. The amounts that are less than the effective dose will most likely not kill the target pest. In this case, the pesticide is applied without the ability to achieve the desired results, that is, elimination of the pest. Instead, this pesticide is added to the environment for no gain. Amounts greater than the effective dose will not necessarily kill the target pest better. Instead, this larger dose may kill more non-target pests, cost more money to apply, and pollute the environment.

Persistence of Pesticides

The half-life is one measure of the persistence of a chemical. The half-life of a substance is the time required for that substance to degrade to one-half its previous concentration. In other words, if a pesticide has a half-life of 10 days, half of the pesticide normally breaks down by 10 days after application. After this time, the pesticide continues to break down at the same rate. In general, the longer the half-life, the greater the potential for movement, simply because it is present in the environment for a longer time. However, the half-life of a material such as a pesticide is not an absolute factor. Soil moisture, temperature, available oxygen, microbial popu lations, soil pH, photo degradation and other factors may cause the half-life of a substance to vary [5].

Table 1: Comparative toxicity of pesticides and natural products

Pesticide	LD_{50} (Rat)/(mg/kg)	Product with almost equal toxicity
TCDD (Dioxin®)	0.0002	Ricin, pure (castor bean extract)
Flocoumafen (Storm®)	0.25	Strychnine
Sarin (GB nerve gas)	0.2	Black widow spider venom
Aldicarb (Temik®)	0.9	Nicotine alkaloid (free base)
Phorate (Thimet®)	1.0	Heroin
Parathion	2.0	Morphine
Carbofuran (Furadan®)	8	Codeine
Nicotine sulphate(Black leaf 40®)	50	Caffeine
Paraquat (Gramoxone®)	150	Benadryl (antihistamine)
Acephate (Orthene®)	833	Salt substitute (KCl)
Allethrin (Pynamin®, Raid®)	1160	Gasoline
Diazinon	1250	Tobacco

Malathion	5500	Caster oil
Ferbam (fungicide)	16900	Mineral oil
Methoprene (Altosid®, Precor®)	34600	Sugar
Pheromones (Checkmate®)	103750	Water

Source: [19, 20]. ®commercial name

Table 2: Characteristics of some commonly used insecticides along with their relative toxicity to fish

Insecticide	Relative run-off potential	Relative leaching potential	Half life in days	Relative **toxicity to fish**[1]
Hydrdamethinon (Amdro®)	large	small	10	high
Diazinon	medium	large	30	high
Chlorpurifos (Durisban®)	large	small	30	very high
Malathion	small	small	1	very high
Acephate (Orthene®)	small	small	3	very low
Carbaryl (Sevin®)	medium	small	10	medium
Dimehoate (Cygon®)	small	medium	7	medium
Trichlorfon (Dylox®)	small	large	27	high

Dicofol (Kethane®)	large	small	60	high
Propargite (Omite®)	large	small	56	high

[1]Fish Toxicity based on catfish and bluegill. LC_{50} categories are rated as follows: very low = more than 100 mg/L, low = 10 to 100 mg/L, medium = 1 to 10 mg/L, high = 0.1 to 1 mg/L, very high = less than 0.1 mg/L. ®commercial name

Acceptable Daily Intake (ADI)

It is used to establish a negligible residue level for pesticide tolerances on human food or animal feed products. This term has been now replaced by another term, negligible residue. Negligible residue means any amount of a pesticide chemical remaining in or on a raw agricultural commodity or group of raw agricultural commodities that would result in a daily intake regarded as toxicologically insignificant on the basis of scientific judgment of adequate safety data [5].

Maximum Contaminant Level (MCL)

This term refers to toxic chemicals regulated as contaminants under the Safe Drinking Water Act (SDWA). Although MCLs do not apply to pesticides specifically, they apply in a general sense. Under SDWA, pesticides are grouped with a larger collection of toxic chemicals that can affect human health when found at certain specific concentrations above established MCLs in drinking water. The Safe Drinking Water Act and the associated regulations try to prevent contamination of drinking water from reaching MCLs through continuous monitoring of water supplies. Regulations under the SDWA establish MCLs in much the same way as FIFRA, FDCA, and the Food Quality Protection Act of 1996 establish pesticide tolerances with negligible residues [5].

PESTICIDES CLASSIFICATION

Insecticides

An insecticide is a pesticide used against insects. They include ovicides and larvicides used against the eggs and larvae of insects respectively. Insecticides are used in agriculture, medicine, industry and the household. The use of insecticides is believed to be one of the major factors behind the increase in agricultural productivity in the 20th century. Nearly all insecticides have the potential to significantly alter ecosystems; many are toxic to humans; and others are concentrated in the food chain. Insecticides applied to crops and in urban areas do not degrade immediately but they break down after a certain period of time. Some of these pesticides are very persistent like organochlorines and remain in the environment for long periods (upto several years). Persistence is a good quality for some pesticides because it means that it remains effective in killing pests for a long time. However, this attribute means that pesticides remain around long enough to enter water sources under some conditions and keep causing toxicity on aquatic organisms for longer durations. Pesticides from the sites of application reach to different water bodies by rainfall and irrigation as they can wash pesticides from areas of application. These pesticides can bioaccumulate in invertebrates and fish species and pass through the food chain to birds, mammals, and finally even to humans.

Herbicides

The extent to which a plant suffers from the effects of a herbicide ranges from extremely little to the plant being highly sensitive, resulting in overall plant death. This range of susceptibility is often referred to as "selectivity". In other words, given herbicides will harm some plant but not others. Some herbicides are referred to as "non-selective" in that they are hazardous to most forms of plant life if applied at dosages recommended for weed control. However, herbicides, work by affecting inherent processes to plants, not mammals or insects. This is the reason for their relatively low order of mammalian toxicity. The persistence of some herbicides can be looked upon as either a detriment or advantage. Obviously, the longer these materials remain active in the soil, the less appealing they are environmentally.

Different herbicides vary widely in their potential to enter water supplies. Some herbicides are water soluble enough to enter into solution with rainfall or irrigation water. Their final destination is highly dependent upon the conditions under which they are applied. They can leach downward or move with the erosion of soil particles if applied to a relatively bare soil surface. The extent to which either of these events occurs depends upon several physical and chemical properties of both the soil and the herbicide.

Fungicides

Fungicides are chemical compounds or biological organisms used to kill or inhibit fungi or fungal spores. Fungi can cause serious damage in agriculture, resulting in critical losses of quality and yield. Fungicides are used both in agriculture as well as to treat fungal infections in animals. Chemicals used to control oomycetes, which are not fungi, are also referred to as fungicides as oomycetes use the same mechanisms as fungi to infect plants. Fungicide can either be contact, translaminar or systemic. Contact fungicides are not taken up into the plant tissue and only protect the plants where the spray is deposited; translaminar fungicides redistribute the fungicide from the upper, sprayed leaf surface to the lower, unsprayed surface. Systemic fungicides are taken up and redistributed through the xylem vessels to the upper parts of the plant. New leaf growth is protected for a short period. Most fungicides are commercially available in a liquid form. The most common active ingredient is sulfur, present at 0.08% in weaker concentrates, and as high as 0.5% for more potent fungicides. Fungicides in powdered form are usually around 90% sulfur and are very toxic. Other active ingredients in fungicides include neem oil, rosemary oil, jojoba oil, and the bacterium Bacillus subtilis. Fungicide residues have been found on food for human consumption, mostly from post-harvest treatments. Some fungicides are dangerous to human health, such as vinclozolin, which has now been removed from use.

Fungicides include as targets a range of pests broader than insecticides. They are an area of concern for maintaining water quality because of their wide use by agriculture and home owners. Fungicides present a clear danger of pollution through their introduction into waters by improper application, storage and disposal. However, additional pollution hazards exist from drift, leaching, and runoff from treated areas where applications have been legal and proper [5].

However, knowing about water pollution potential of fungicides, one can plan their use and minimize chances of these chemicals entering surface and groundwater. Fungicides work in a variety of ways. The ability of the target organisms to rapidly develop resistance has generated a wide variety of chemical actions. The persistence of some fungicides offers advantages and disadvantages to both the user and the environment. The more persistent fungicides present the hazard of remaining in the environment long enough to enter soil and water profiles. It is also important that the fungicide active ingredient may not be as toxic or as environmentally hazardous as some of the inert ingredients in the formulation [5].

ENTRY OF PESTICIDES INTO WATER SYSTEMS

Cook, et al. [5] had mentioned that pesticides can enter water through surface runoff or through leaching. These two fundamental processes are linked to the earth's hydrologic cycle. When we include urban water use in surface runoff, pesticide residues in municipal wastewater fit the hydrologic model. Figure 3 shows the hydrologic cycle and gives a graphic representation of the various routes water takes to reach a low point. When water enters an established body of water or backs-up behind a barrier, it carries with it the dissolved materials that it picked up in the media through which it flowed. Figure 4 shows the routes pesticide pollutants may take to reach surface or groundwater. It is difficult to determine how materials that become water pollutants actually get into water sources. Often it is the action of water itself that causes pollutants to enter bodies of water. The source of water that transports pollutants may be natural, such as rainfall, or caused by humans, as in the case of irrigation or diversion of water. Pollutants also may enter bodies of water by wind or by their own passive movement. Movement of pollutants is a complex system and pesticides can come from either point sources or nonpoint sources. Point sources are small, easily identified objects or areas of high pesticide concentration such as tanks, containers, or spills. Non-point sources are broad, undefined areas in which pesticide residues are present.

Figure 3: The hydrologic cycle. Different levels of water evaporation are shown. Water always flows in the lowest point.(Source: texas agriculture extension service, the texas A&M university system, "pesticide characteristics that effect water quality", Jerry L.Cook, Paul Baumann, John A Jackmang and Doung Stevenson, texas A&M university, college

station, TX 77842

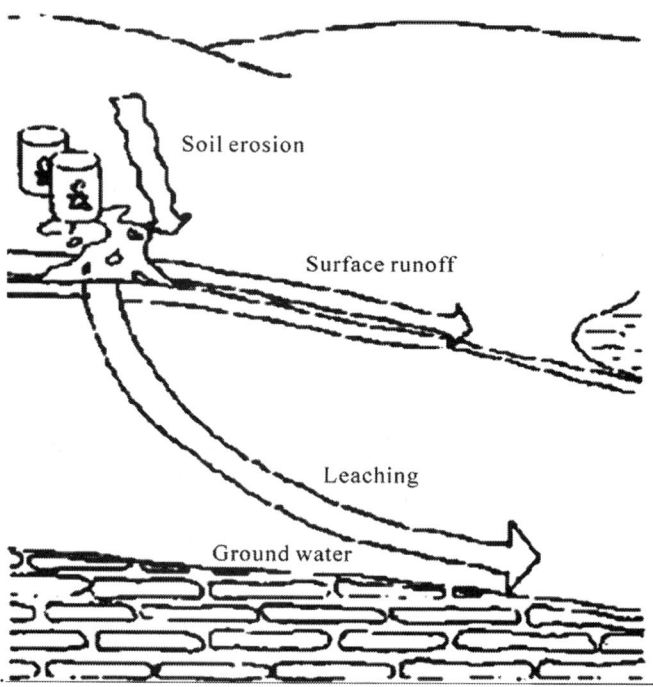

Figure 4: The pesticides can pollute water either though soil erosion, surface runoff or leaching. (Source: Texas agriculture extension service ,the Texas A&M university system,

Surface Runoff

Water that flows across the surface, whether from rain, irrigation, or other water released onto the surface, always flows downhill until it meets with a barrier, a body of water, or begins to percolate into the soil. Figures 3 and 4 show examples of surface runoff and how it can pick up and carry pesticides into surface or groundwater [5].

Agriculture and Water Quality Conflicts

Water quality problems, thought to be caused in part by cropland runoff or non-point source pollution, affect drinking water and the nation's lakes, streams, and estuaries. Action taken by public officials to protect our water resources may change the diversity, quality, and quantity of farm products, production systems, and ultimately the prices consumers pay. Losses from impaired water quality can cost billions of dollars, not just to agriculture but also to recreation, commercial fishing, municipal water treatment, and river navigation.

Groundwater resources are vulnerable to contamination from many directions (Figure 2). When animal wastes, fertilizers, herbicides, insecticides, and fungicides are applied to cropland, some residues remain in the soil after plant uptake and may leach into subsurface waters, or the residues may move to surface water by dissolving in runoff or adsorbing to sediment. Spray drifts during application may carry pesticides to surface waters also. Chemical or physical processes transform residues into products that may also contaminate water. For example, nitrogen fertilizer or nitrogen from animal waste may be transformed first into ammonium and then into nitrates. Nitrates can turn into nitrites and both are detrimental to human health.

Nutrients, particularly nitrogen and phosphorus from fertilizers, promote algae growth and premature aging of lakes, streams, and estuaries (a process called eutrophication). Suspended sediment impairs aquatic life by reducing sunlight, damaging spawning grounds, and may be toxic to aquatic organisms. Pesticide residues that reach surface water systems may also affect the health and vigor of freshwater and marine organisms.

WATER QUALITY CONCERNS

Drinking water for humans from contaminated wells is exposed to pesticide and fertilizer residues. A known human health risk from nitrate contamination is infant methemoglobinemia, a condition where nitrates are converted into nitrites in the digestive system, impairing the ability of infants' blood to carry oxygen. Nitrites are also considered carcinogenic (tumor causing) by some analysts. Concentration of nitrates or pesticides in drinking water may be below levels at which acute health effects have been observed. However, continued exposure may result in chronic effects (i.e., reproductive impairments, cancer, etc.) to humans or other organisms. The degree of health risk associated with drinking water containing traces of pesticides or nitrates at, or below, levels where human health could be endangered is poorly understood.

Some pesticides are considered carcinogenic in large doses, and as a result, the United States Environmental Protection Agency (EPA) has issued health standards defining maximum allowable contamination levels for 26 pesticides. Contaminated groundwater that resurfaces also affects nontargeted plants, birds, or aquatic organisms (some of which are endangered) in the environment. Due to several years of control efforts, the share of pollution from point sources, such as discharges from sewage treatment plants or industrial sources, appears to be lessening. According to the EPA, the non-point source pollution resulting from agricultural tillage, pesticide application, and urban development sites is the chief cause of surface water degradation today.

Agricultural runoff is the single most extensive source of surface water pollution, accounting for 55 percent of impaired river miles and 58 percent of impaired lake acres assessed by the States in 1986 and 1987. In a recent study by USDA's Economic Research Service (ERS), the degree to which agricultural runoff contributed to delivery of nutrients and sediments to lakes and streams were calculated. Out of 99 watersheds examined, 48 had excessive levels of nutrients or sediment. The study found agriculture to be a "significant source" (defined as contributing more than 50 percent of pollutant discharge) of nitrogen in nine watersheds. Agricultural sources of sediment were significant in 34 watersheds. Thirty-one watersheds had significant agricultural discharge of phosphorus. Another recent ERS study identified the scope and significance of agricultural contributions to coastal water pollution. For the 78 estuarine systems considered, agricultural runoff supplied an average of 24 percent of total nutrients and 40 percent of total sediment. Agriculture contributed more than 25 percent of total nutrients in 22 of the 78 estuaries. High rates of pesticide losses to surface waters were found in 21 systems. Fifteen estuarine systems showed both significant agricultural nutrients and high pesticide losses.

The extent to which the nation's groundwater resources are affected by agricultural chemicals is less well known. Discoveries of chemical residuals in groundwater during the late 1970's and early 1980's dispelled the commonly held view that groundwater was protected from agricultural chemicals by impervious layers of rock, soil, and clay. Groundwater may also be contaminated by other sources, including nonagricultural use of pesticides and fertilizers, and leaking underground storage tanks.

FACTORS AFFECTING PESTICIDE TOXICITY IN AQUATIC SYSTEMS

The ecological impacts of pesticides in water are determined by the following criteria:

Toxicity

Mammalian and non-mammalian toxicity usually expressed as LD_{50} ("Lethal Dose"): The lower the LD_{50}, the greater the toxicity; values of 0-10 are extremely toxic [22]. Drinking water and food guidelines are determined using a riskbased assessment. Generally, Risk = Exposure (amount and/or duration) × Toxicity. Toxic response (effect) can be acute (death) or chronic (an effect that does not cause death over the test period but which causes observable effects in the test organism such as cancers and tumours, reproductive failure, growth inhibition, teratogenic effects, etc.).

Persistence

Measured as half-life is determined by biotic and abiotic degradational processes. Biotic processes are biodegradation and metabolism; abiotic processes are mainly hydrolysis, photolysis, and oxidation [23]. Modern pesticides tend to have short half-lives that reflect the period over which the pest needs to be controlled.

Degradates

The degradational process may lead to formation of "degradates" which may have greater, equal or lesser toxicity than the parent compound. As an example, DDT degrades to DDD and DDE.

Fate (Environmental)

The environmental fate (behaviour) of a pesticide is affected by the natural affinity of the chemical for one of four environmental compartments [23]: solid matter (mineral matter and particulate organic carbon), liquid (solubility in surface and soil water), gaseous form (volatilization), and biota. This behaviour is often referred to as "partitioning" and involves, respectively, the determination of: the soil sorption coefficient (K_{OC}); solubility; Henry's Constant (H); and the n-octanol/water partition coefficient (K_{OW}). These parameters are well known for pesticides and are used to predict the environmental fate of the pesticide. An additional factor can be the presence of impurities in the pesticide formulation but that are not part of the active ingredient. A recent example is the case of TFM, a lampricide used in tributaries of the Great Lakes for many years for the control of the sea lamprey.

EFFECTS OF PESTICIDES ON HUMAN HEALTH

Perhaps the largest regional example of pesticide contamination and human health is that of the Aral Sea region. UNEP (1993) [24] linked the effects of pesticides to "the level of oncological (cancer), pulmonary and haematological morbidity, as well as on inborn deformities and immune system deficiencies". Human health effects are caused by 1) Skin contact: handling of pesticide products, 2) Inhalation: breathing of dust or spray and 3) Ingestion: pesticides consumed as a contaminant on/in food or in water. Farm workers have special risks associated with inhalation and skin contact during preparation and application of pesticides to crops. However, for the majority of the population, a principal source is through ingestion of food which is contaminated by pesticides. Degradation of water quality by pesticide runoff has two principal human health impacts. The first is the consumption of fish and shellfish that are contaminated by pesticides; this can be a particular problem for subsistence fish economies that lie downstream of major agricultural areas. The second is the direct consumption of pesticide-contaminated water. WHO (1993) [25] has established drinking water guidelines for 33 pesticides. Many health and environmental protection agencies have established "acceptable daily intake" (ADI) values that indicate the maximum allowable pesticide daily ingestion over a person's lifetime without appreciable risk to the individual. For example, Wang and Lin (1995) [26] studying substituted phenols, tetrachlorohydroquinone, a toxic metabolite of the biocide pentachlorophenol, was found to produce significant and dose-dependent DNA damage. The harmful efects of pesticides are 1) Death of the organism, 2) Cancers, tumours and lesions on fish and animals, 3) Reproductive inhibition or failure, 4) Suppression of immune system, 5) Disruption of endocrine (hormonal) system, 6) Cellular and DNA damage, 7) Teratogenic effects (physical deformities such as hooked beaks on birds), 8) Poor fish health marked by low red to white blood cell ratio, excessive slime on fish scales and gills, etc., 9) Intergenerational effects (effects are not apparent until subsequent generations of the organism) and 10) Other physiological effects such as egg shell thinning. These effects are not necessarily caused solely by exposure to pesticides or other organic contaminants, but may be associated with a combination of environmental stresses such as eutrophication and pathogens [27,28].

Pesticides are commonly found in water. The groundwater from some US and Canadian provinces has been reported to contain the residues of 39 pesticides and their metabolites [29]. The calculation of level of allowable pesticide for water is made depending on the exposure of children and adults exposure; the children being 4 times more vulnerable to the pesticide toxicity than adults [30]. Residues of pesticides that are "severely restricted" because of their serious effects on human health were also found in significant quantities in the water sources. The pesticide residues exerting serious effects on human health enter the water supply through leaching from soil into ground water.

ABSORPTION OF PESTICIDES THROUGH SKIN AND RESPIRATORY ROUTES

The reports available indicate that the infants and children absorb more pesticides and their residues, insect repellents and pediculocides than the adults through their skin and produce toxicity [29]. It leads to alterations in behavioural pattern and several diseases syndromes such as encephalopathy, ataxia, seizures, muscle cramps, frequent urination and coma [30,31]. However, farmers generally get exposed to the pesticides via spraying of these chemicals into the fields. The absorption of pesticides in farmers through cutaneous and respiratory routes predominantly contributes to the overall pesticide toxicity in them which has been reported to cause non-Hodgkins lymphoma [32].

REMOVAL OF TOXIC SUBSTANCES INCLUDING ARSENIC FROM DRINKING WATER

Reverse Osmosis (RO) is a process to get rid of all the impurities in drinking water including deadly ions and organisms and pesticide/fertilizer residues. Under RO systems, water is made to pass through a membrane having a pore size of 0.0001 micron under high pressure. Only 5-10 percent of the ions are able to slip across the membrane, which is well within acceptable levels as per all standards including WHO, BIS, etc. RO systems are suitable for removing several of the toxic substances present in water in dissolved form, including fluoride, fertilizer and pesticide residues, and heavy metals. But costs vary, depending on the plant capacity and level of utilization, the level of salinity and other impurities in the water and the distance from the source of water. Costs can range between Rs. 0.03/litre (for brackish water) to Rs. 0.10/litre (for seawater).

A household arsenic treatment method is the ferric chloride coagulation system. This involves precipitation of arsenic by adding a packet of coagulant in 25 litres of tube well water, and subsequent filtration of the water through a sand filter. Field experiments showed arsenic concentration in treated water was nearly 1/20 that of raw water. The cost of chemical (ferric chloride) for treatment is Rs. 0.09/litre of raw water to be treated.

Another method for removing arsenic is based on "sorptive filtration based on iron coated sand bed". Water is first put in a bucket and stirred for some time to accelerate precipitation of excess iron. It is then allowed to pass through a sand filter where the excess iron is filtered out. Finally the water is passed through an iron coated sand filter. But, the efficiency of removing arsenic reduces drastically beyond a certain bed volume with the arsenic concentration of treated water crossing the permissible limit of 50 ppb. The third method involves filtration of arsenic from raw water by passing it through a gravel media containing iron sludge. An evaluative study showed the first two systems to be superior, with the first one found to be most acceptable to the villagers.

WATER POLLUTION CASE STUDY SHOWS PESTICIDE POLLUTION IN INDIA

One of the most terrifying effects of pesticide contamination of groundwater came to light when pesticide residues were found in bottled water. Between July and December 2002, the Pollution Monitoring Laboratory of the New Delhi-Based Centre for Science and Environment (CSE) analyzed 17 brands of bottled water; both packaged drinking water and packaged natural mineral water, commonly sold in areas that fall within the national capital region of Delhi. Pesticide residues of organochlorine and organophosphorus pesticides, which are most commonly used in India, were found in all the samples. Among the organochlorines, gamma-hexachlorocyclohexane (lindane) and DDT were prevalent, while among organophosphorus pesticides, Malathion and Chlorpyrifos were the most common. All these were present above the permissible limits specified by the European Economic Community (EEC), which is the norm, used all over Europe. One may wonder as to how these pesticide residues get into bottled water that is manufactured by several big companies. This can be due to several reasons. There is no regulation that the bottled water Industry must be located in 'clean' zones. Currently, the manufacturing plants of most brands are situated in the dirtiest industrial estates or in the midst of agricultural fields. Most companies use bore-wells to pump out water from the ground from depths varying from 24-152 m below the ground. The raw water samples collected from the plants also revealed the presence of pesticide residues.

Thus, the fault obviously lies in the treatment methods used. These plants use membrane technology, where the water is filtered using membrane with ultra-small pores to remove fine suspended solids and all bacteria and protozoa and even viruses. While nanofiltration can remove insecticides and herbicides but it is expensive and thus rarely used. Most industries also use an activated charcoal adsorption process, which is effective in removing organic pesticides but not heavy metals.

To remove pesticides, the plants use reverse osmosis and granular activated charcoal methods. So even though the manufacturers claim to use these processes, the presence of pesticide residues points to the fact that either the manufacturers do not use the treatment process effecttively or only treat a part of the raw water. The low concentrations of pesticide residues in bottled water do not cause acute or immediate effects.

However, repeated exposure even to extremely miniscule amounts can result in chronic effects like cancer, liver and kidney damage, disorders of the nervous system, damage to the immune system and birth defects. CSE reported pesticide residues in bottled water as well as in popular cold drink brands sold across the country. This is because the main ingredient in a cold drink or a carbonated non-alcoholic beverage is water and there are no standards specified for water to be used in these beverages in India. There were no standards for bottled water in India till September 29, 2000, when the Union Ministry of Health and Family Welfare issued a notification (No. 759(E)) amending the Prevention of Food Adulteration Rules, 1954. The BIS (Bureau of Indian Standards) certification mark became mandatory for bottled water from March 29, 2001. However, the parameters for pesticide residues remained ambiguous. A series of Committees were established and eventually on 18th July 2003, amendments were made in the Prevention of Food Adulteration Rules stating that pesticide residues considered individually should not exceed 0.0001 mg/L and that the total pesticide residues should not be more than 0.0005 mg/L and that the analysis shall be conducted by using internationally established test methods meeting the residue limits specified herein. This notification came into force from January 1, 2004.

NEW SYSTEM MINIMIZES PESTICIDE POLLUTION IN AQUIFERS

The recent report from the Institute of Natural Resources and Agrobiology of the Spanish National Research Council (CSIC) has suggested the development of a new method to encapsulate and slowly release pesticides to prevent the leaching as well as the volatilization of these molecules. This new method helps to encapsulate the pesticide in lecithin liposomes or vesicles leading to the adsorption on clay. The final complex comprising liposomes, pesticide and clay is dispersed in water, allows the chemical compound to be slowly released. This entrapment technique restricts the spread of pesticides and their residues to other surfaces and aquifers; thereby acting as substances of minimal toxicological concern [33].

POLLUTED RIVER STRETCHES

Environmental Factors Influencing River Water Quality

Due to environmental conditions such as basin lithology, vegetation and climate the river water quality varies. In small watershed, spatial variations extends over orders of magnitude for most major elements and nutrients, while this variability is of lesser magnitude for lower major basins. Therefore, the standard river water use for reference is not applicable and it is because of this that the natural waters can possibly is unfit for various human uses, even including drinking.

The rivers carry three major natural sources of dissolved and soluble matter namely the atmospheric inputs of material, the degradation of terrestrial organic matter and the weathering of surface rocks. These substances are carried through soil and porous rocks and finally reach the rivers. On their way, they are affected by numerous processes such as recycling in terrestrial biota and storage in soils. The exchange between dissolved and particulate matter and loss of volatile substances to the atmosphere, production and degradation of aquatic plants within rivers and lakes etc. get affected. As a result of these multiple sources and pathways, the concentrations of elements and compounds found in rivers depend on physical factors (climate, relief), chemical factors (solubility of minerals) and biological factors (uptake by vegetation, degradation by bacteria). The most important environmental factors controlling river chemistry are 1) Occurrence of highly soluble (halite, gypsum) or easily weathered (calcite, dolomite, pyrite, olivine) minerals, 2) Distance to the marine environment which controls the exponential decrease of ocean aerosols input to land

(Na^+, Cl^-, SO_4^{2-}, and Mg^{2+}), 3) Aridity (precipitation/runoff ratio) which determines the concentration of dissolved substances resulting from the two previous processes, 4) Terrestrial primary productivity which governs the release of nutrients (C, N, Si, K), 5) Ambient temperature which controls, together with biological soil activity, the weathering reaction kinetics and 6) Uplift rates (tectonism, relief) Stream quality of unpolluted waters (basins without any direct pollution sources such as dwellings, roads, farming, mining etc.

River Water Pollution

Most of the Indian rivers and their tributaries viz., Ganges, Yamuna, Godavari, Krishna, Sone, Cauvery Damodar and Brahmaputra are reported to be grossly polluted due to discharge of untreated sewage disposal and industrial effluents directly into the rivers. These wastes usually contain a wide variety of organic and inorganic pollutants including solvents, oils, grease, plastics, plasticizers, phenols, heavy metals, pesticides and suspended solids. The indiscriminate dumping and release of wastes containing the above mentioned hazardous substances into rivers might lead to environmental disturbance which could be considered as a potential source of stress to biotic community. As for example, River Ganges alone receives sewage of 29 Class I cities situated on its banks and the industrial effluents of about 300 small, medium, and big industrial units throughout its whole course of approximately 2525 km. Identically Yamuna is another major river, has also been threatened with pollution in Delhi and Ghaziabad area. Approximately 515,000 kilolitres of sewage waste water is reported to be discharged in the river Yamuna daily. In addition, there arc about 1,500 medium and small Industrial units which also contribute huge amounts of untreated or partially treated effluent to the river Yamuna every day.

Similarly many other rivers were surveyed during past two decades with respect to their pollutional status. In addition to domestic and industrial discharge into the rivers, there were continued surface run off of agricultural areas, mines and even from cremation on the river banks. According to a report, over 32 thousand dead bodies were cremated at the major burning Ghats per year in Varanasi alone in the year 1984.

Pollution in the Ganga River

The Ganga Basin, the largest river basin of the country, houses about 40 percent of population of India. During the course of its journey, municipal sewages from 29 Class I cities (cities with population over 100,000), 23 Class II cities (cities with population between 50,000 and 100,000) and about 48 towns, effluents from industries and polluting wastes from several other non-point sources are discharged into the river Ganga resulting in its pollution. The NRCD records, as mentioned in audit report, put the estimates of total sewage generation in towns along river Ganga and its tributaries as 5044 MLD (Million Litres per Day). According to the Central Pollution Control Board Report of 2001, the total wastewater generation on the Ganga basin is about 6440 MLD.

Many towns on the bank of the Ganga are highly industrialised. Most of the industries have inadequate effluent treatment facilities and dump their wastes directly into the river. A high concentration of tanneries in Kanpur has further aggravated the situation. Besides other chemical and textile industries, Kanpur has 151 tanneries located in a cluster at Jajmau along the southern bank of the Ganga with an estimated waste water discharge of 5.8 to 8.8 million litres per day. Out of 151 tanneries in Jajmau, 62 tanneries use exclusively the chrome tanning process, 50 tanneries use vegetable tanning processes, and 38 tanneries use both chrome and vegetable tanning. The Indian government under the Ganga Action Plan (GAP) has implemented several schemes for the abatement of pollution of the Ganga by tanneries. However, there are violations of the pollution control measures, and tannery effluents are still found in the river.

Pollution in the Yamuna River

River Yamuna is the primary source of drinking water for Delhi, the capital of India, and also for many cities, towns and villages in the neighbouring states of Uttar Pradesh, Uttaranchal and Haryana. In the last few decades, however, there has been a serious concern over the deterioration in its water quality. The river has been receiving large amounts of partially treated and untreated wastewater during its course, especially between Wazirabad and Okhla, National Capital Territory

(NCT) of Delhi. Pollutants flowing into the river are contributed from the waste of the cities situated along its bank. Once the lifeline of Delhi, Yamuna has now become the most polluted water resource of the country. It now looks like a sewer. From big industries and factories to people living in big colonies, slums and rural areas, all pollute the river with impunity because of untreated water. Increasing pollution of the Yamuna has now become an international issue and a cause of concern for environmentalists.

Impact of River Water Pollution

The pollutants include oils, greases, plastics, plasticizers, metallic wastes, suspended solids, phenols, toxins, acids, salts, dyes, cyanides, pesticides etc. Many of these pollutants are not easily susceptible to degradation and thus cause serious pollution problems. Contamination of ground water and fish-kill episodes are the major effects of the toxic discharges from industries. The impact involves gross changes in water quality viz. reduction in dissolved oxygen and reduction in light penetration that's tends loss in self purification capability of river water.

On the worldwide scale, the river water pollution leads hazardous impact on aquatic animals and plants. Some studies show alarming condition of river pollution implications. Pratap and Vandana [34] performed detailed study on pesticide accumulation in Fish species and concluded that, pesticide bioaccumulation was higher in cat fishes as compared to carps and have species specific in their tissues (liver, brain and ovary) causing metabolic and hormonal imbalance affecting at GnRH and GTH secretion. The reproductive sex steroid hormones were lowered in cat fishes and carps of the polluted rivers. They suggested that the bioaccumulated insecticide in ovary may cause blocking of the receptor site so that natural hormone cannot bind at the site of estrogen receptor which may cause the dysfunctions of the reproduction in cat fishes and carps inhabiting the polluted river Gomti and Ganga. They also suggested that the fish bioaccumulated insecticide beyond permissible limit must be avoided for the food purpose from such polluted rivers.

Contamination by synthetic organic pollutants is a more recent phenomenon which is even more difficult to demonstrate for lack of appropriate monitoring. The DDT content of the Yamuna River which flows through Delhi is one of the highest ever reported, many other problems affect river water quality on a global scale. Very severe pollution by pathogenic microorganisms is still the prime cause of waterborne morbidity and mortality although it is difficult to establish reliable statistical correlation in each case. Many streams and rivers in South America, Africa and particularly on the Indian sub-continent show high coliform levels together with high BOD and nutrient levels. Eutrophication, which has spread widely to lakes and reservoirs of developing countries now also, affects slow flowing rivers.

Another shocking incident came in picture recently, shows a death alarm of river pollution. Yamuna river water is behind death of crocodiles in the Chambal Sanctuary. Chambal lost over 100 crocodiles in the last 72 days to a mysterious toxin released, in all possibility, by its very own sanctuary—the river Yamuna. Initially crocodile's deaths were reported from 35 km stretch of National Chambal Sanctuary, where the Chambal and Yamuna rivers meet, but now crocodile's deaths are reported from upstream also. Beside, other forms of aquatic life are also coming in the area of the impact. For instance, two dolphins and a Crocodile have also died recently. Vets and research labs involved in the probe have confirmed that toxins caused around 103 deaths. They unanimously agree toxins came from either the contaminated food or the Yamuna water. After almost three months since 16 bodies were fished out from Barchauli village in Etawah range of national Chambal sanctuary on December 8, it is gout which has been noted in regularity in all 103 carcasses. The bodies show uric acid deposition in visceral organs and also joints of animals. Initial findings point towards ecological degradation of river system. Experts agree that Tilapia, an exotic fish species, could be the possible carrier of toxins and consumption of this species by crocodiles may have led to their death.

Prevention and Control of Pollution

Some actions have been taken by The Government of India to control pollution in the river systems. Ganga action plan is much known of them. Ganga Action Plan (GAP) was launched for immediate reduction of pollution load on the river Ganga. It was prepared by Department of Environment (now Ministry of Environment & Forests) in December 1984 on the basis of a survey on Ganga basin carried out by the Central Pollution Control Board in 1984. The Plan approved by the Government in April 1985 pursued two objectives: to reduce the pollution load in the Ganga and establish sewage treatment systems in 25 Class I cities bordering the river. To oversee the implementation of the GAP and lay down policies and programmes, Government of India constituted the Central Ganga Authority (CGA) under the chairmanship of the Prime Minister in February 1985. It has been renamed as the National River Conservation Authority (NRCA) in September 1995, as a wing of the Department of Environment, to execute the projects under the guidance and supervision of the CGA. The state agencies like Public Health Engineering Department, Water and Sewage Boards, Pollution Control Boards, Development Authorities, Local Bodies etc. were responsible for actual implementation of the scheme.

Failure of Ganga Action Plan

The Ganga action plan launched by the Government of India with much fanfare has failed in achieving its objectives. The pollution levels in Ganga are either same or even higher. The Sankat Mochan Foundation found that the schemes for Varanasi-India under the GAP Phase-I suffered from several shortcomings. Some major ones are 1) The sewage pump at Konia terminal, when run to its capacity causes heavy surcharging of the old trunk sewer. It causes erosion of the sewer linings and also spillage of sewage from manholes in low-lying areas of the city, 2) Over 115 MLD sewage, which could be easily handled by the Konia Terminal, is actually being diverted to Dinapur Sewage Treatment Plant. The Dinapur STP can handle only 80 MLD, resulting in by-passing of 35 MLD untreated sewage into Varuna and eventually into Ganga. This is also very expensive in terms of energy consumption, 3) Power breakdowns, which are common in Varanasi, causes a sudden back pressure in the system and massive spillage of sewage onto the roads and streets of the city, 4) The plant at Dimapur has to be shut down completely during monsoons. Thus for three to four months in a year all the sewage goes untreated, 5) The biogas generator in the Dinapur STP does not function hence the plant is ineffective due to shortage of power. Tens of millions of Rupees have been wasted on its construction, while the villages around the Dinapur STP suffer from polluted water, water borne diseases and mosquitoes.

The observations on the GAP Phase I schemes for Varanasi-India indicate that: 1) BOD in the religious bathing area remains dangerously high even after completion of the GAP I. The BOD is as high as 25 mg/L at the confluence of Ganga and Varuna rivers. 2) The faecal coliform varied from 70000 mpn/100mL to 1.5 million mpn/100mL. The BOD and the faecal coliform levels increase from upstream to downstream as more and more untreated sewage enters the river. 3) These values when compared with those six km upstream of Assi are an eye opener. The figures in this area, where the city of Varanasi starts and no point discharges of effluents take place are 2 mg/L of BOD and undetectable faecal coliform. 4) Even in the treated sewage coming out from the Dimapur STP, the BOD is dangerously high at 50mg/L against a maximum permissible value of 20mg/L. Suspended solids are 100mg/L. Faecal coliform levels remain as high as that entering the STP, since there is no arrangement for controlling it.

According to environmentalists, about 90 percent of pollution into the holy river is caused by sewage generation while only about 5 to 6 percent can be blamed on bathing and other activities. The real sources of pollution i.e. sewage, however, still continues to flow into the river. By 1996, the first phase of the Ganga Action Plan (GAP) was completed and the government expanded its pollution abatement activities by enlarging the bureaucracy. They created the National River Conservation Directorate (NRCD) and folded the (GAP) into that Directorate.

The Ganga Action Plan Phase II

Since GAP I did not cover the pollution load of Ganga fully; the Ganga Action Plan Phase II (GAP II) was launched in stages between 1993 and 1996. 1) On the tributaries of river Ganga viz. Yamuna, Damodar and Gomati. 2) In 25 class-I towns left out in Phase I. 3) In the other polluting towns along the river. The Cabinet Committee on Economic Affairs (CCEA) approved the GAP-II in various stages during April 1993 to October 1996. The States of Uttarakhand, Haryana, Delhi, Uttar Pradesh, Bihar and West Bengal were to implement the GAP-II by treating 1912 MLD of sewage. Against this, a treatment capacity of 780 MLD has been created so far (October 2003). The approved cost of GAP II is Rs. 22854.8 million (excluding establishment charges) against which, an amount of Rs.7923.8 million has been released till 30 November 2003. The total number of schemes sanctioned under GAP II so far is 495 at a cost of Rs.13800 million, out of which 318 schemes have been completed. The revised date for completion of GAP II was kept as December 2005. The Ministry of Environment and Forests have now stated that as the second Phase of Gomti Action Plan and Yamuna Action Plan had been approved and these were targeted to be completed by March 2007 and September 2008, respectively, the GAP II was targeted to be completed by December 2008, subject to the availability of funds in time.

Water Pollution—Related Legislation

The first significant law regarding the protection of environmental resources appeared in the 1970's with the setting up of a National Committee on Environmental Planning and Coordination, and the enactment of the Wildlife Protection Act, 1972. Since then, three main texts have been passed at the central level that is relevant to water pollution: the Water (Prevention and Control of Pollution) Act, 1974, the Water (Prevention and Control of Pollution) Cess Act, 1977 and the Environment (Protection) Act (1986). The Water Act 1974 established the Pollution Control Boards at the central and state level. The Water Cess Act 1977 provided the Pollution Control Boards with a funding tool, enabling them to charge the water user with a cess designed as a financial support for the board's activities. The Environment Protection Act 1986 is an umbrella legislation providing a single focus in the country for the protection of environment and seeks to plug the loopholes of earlier legislation relating to environment. The law prohibits the pollution of water bodies and requires any potentially polluting activity to get the consent of the local SPCB before being started.

Use of Informal Regulation of Pollution

The design of policy instruments for industrial pollution is not only complex but also very daunting in the case of developing countries. In principle, the regulator has an array of physical, legal, monetary, and other instruments at his/her disposal. But the presence of a large number of pollution sources in the form of small-scale industries (SSIs) that lack knowledge, funds, technology and skills to treat their effluent frustrates any instrument applied and leads to overall failure. The failure of industrial pollution control is also attributable to rigid commandand-control regulatory approaches. Regulators are constrained by meagre resources, limited authority and political interference. These problems are compounded by information asymmetries. For all these reasons, numerous studies in India have concluded that despite a strong legal framework and the existence of a large bureaucracy to manage environmental regulation, implementation is very weak. The failure of formal regulation to control pollution has highlighted the significance of informal regulation for achieving environmental goals. There is now considerable interest in "information disclosure" and "rating" as potential tools of industrial pollution control. Sometimes referred to as the "third wave" of environmental policy, this approach acknowledges the difficulties of monitoring and enforcement and recognises that there are many more avenues of influence than just formal regulation or fines. Firms are sensitive, for example, about their reputation and the future costs that they may incur as a result of liability or accidents. The emergence of this new paradigm for regulation is also related to advances made in our understanding of asymmetric information. Goldar and Banerjee [35] made an attempt to assess the impact of informal regulation of pollution on water quality in Indian rivers. For this purpose, an econometric analysis of determinants of water quality in Indian rivers were carried out using water quality (water class) data for 106 monitoring points on 10 important rivers for five years, 1995-1999. Results showed significant favorable effect of informal regulation of pollution on water quality in rivers in India.

The water quality data generated through National Water Monitoring Programme and River Basin Studies carried out since 1980 indicated deterioration of water quality in riverine segments and other water bodies. The water bodies not meeting the desired water quality criteria are identified as polluted river stretches/water bodies. The deviation of water quality from the desired water quality criteria in the data generated for the river Ganga formed the basis for launching Ganga Action Plan (GAP). Subsequently, 10 river stretches not meeting the desired criteria were identified during 1988-1989. The list of polluted stretches increased to 37 during the year 1992 covering all the major river basins. The polluted river stretches were intensively surveyed by Central Pollution Control Boards (CPCB) and State Pollution Control Boards (SPCBs) to identify the sources of pollution such as Urban Centres and Industrial Units. With the expansion of monitoring network and coverage of more number of rivers for regular monitoring, the numbers of polluted water bodies identified during 2002 are 86 (71 rivers and 15 lakes/ponds/creeks), which are not meeting the desired criteria. Statewise number of polluted stretches in rivers and lakes is given in Table 3.

What is happening to the Yamuna is reflective of what is happening in almost every river in India. More than 700 million Indians do not have adequate sanitation. The United Nations says that 2.1 million children under 5 die each year because of a lack of clean water and the World Bank has warned India that it stands on the edge of an era of severe water scarcity. Nothing illustrates this more vividly than the Yamuna. The Government extracts 1.1 billion litres from it daily, making it the capital's largest water source. By the time the river leaves Delhi, it turns into a vast drain, carrying an estimated 3.5 billion litres of sewage every day. Its oily black waters cannot sustain fish or plant life. Methane bubbling from its surface can be smelt across the city. Since 1992 the Government has spent 20 billion rupees (£240 million) on cleaning the river but it is not visible. Pollution levels have doubled and less than half of the sewage in the river is treated.

Table 3: State wise polluted stretches in rivers and lakes in India

Name of State	No. of Water Bodies	River	Lake/Tank/ Drain etc.
Andhra Pradesh	8	3	5
Assam	2	2	
Delhi	1	1	
Jharkhand	1	1	
Gujarat	10	9	1

Haryana	3	2	1
Himachal Pradesh	2	1	1
Karnataka	6	4	2
Madhya Pradesh	5	4	1
Maharashtra	15	15	
Meghalaya	5	1	4
Orissa	5	5	
Punjab	3	3	
Rajasthan	3	3	
Tamil Nadu	7	7	
Sikkim	1	1	
Uttar Pradesh	8	8	
West Bengal	1	1	
TOTAL: -	86	71	15

Source: Water pollution (Polluted river stretches) [35]

BHOPAL GAS TRAGEDY

Bhopal's pesticide plant was built in 1969 to manufacture Sevin, a pesticide used throughout Asia to kill beetles, weevils and worms. The plant was operated by Union Carbide India, Limited, but an American company, Union Carbide Corporation, held more than half the stock. The leak began on December 2, 1984, when water entered a tank that was used to store methyl isocyanate, a toxic gas and a key ingredient in Sevin. The water reacted with the gas, causing extreme pressure and heat that possibly caused the tank to explode. The tank spewed 40 tons of poisonous gas into the air. The toxic cloud was mostly methyl isocyanate, a compound that can irritate the throat and eyes, cause chest pain and shortness of breath, and, in large doses trigger convulsions, lung failure and cardiac arrest. It is also presumed that the reactions inside the tank generated enough heat to turn methyl isocyanate into its even deadlier cousin: hydrogen cyanide. Listed as a chemical weapon by the Chemical Weapons Convention, hydrogen cyanide can stop respiration. Because the deadly mixture was heavier than the air, it stuck close to the ground, choking thousands of people who lived nearby. The areas engulfed by gas were some of Bhopal's poorest neighborhoods. Many of the gas survivors are still too ill to work and a number of additional health problems continue to crop up such as blindness, respiratory illnesses, reproductive problems and neurological and immune disorders, to name some of them. Due to the rains, the plant's waste ended up in the groundwater. In 1996, the state pollution control board found traces of pesticides in the local wells. But it wasn't until 2004 that the federal government ordered the state to provide the community with clean drinking water. After understanding the causes, it becomes necessary to know the consequences of water pollution. All the water pollutants are responsible for decreasing the selfpurifying ability of the water bodies. This means that these lose the capacity to recycle the wastes. Nutrients cause excessive weed growth and algal blooms, which may release the algal toxins like microcystins and other hazardous compounds.

REPERCUSSIONS OF WATER POLLUTION

The repercussions of this issue are many. Water clarity is affected and the water bodies become shallower. Algae consume most of the available oxygen, thereby increasing what is termed as the Biological Oxygen Demand (BOD) and decreasing the Dissolved Oxygen (DO) level. Also, the rate of photosynthesis is decreased, killing many aquatic plants. Soil erosion brings a lot of silt into the water bodies, thus decreasing the water quality. The lying of cow dung along the periphery of water bodies enriches them with undesirable chemicals. Water pollution as such leads to water borne diseases like cholera, typhoid, diarrhea, hepatitis, jaundice, dysentery etc. Various unwanted plants and effluents give them a marsh-like look, not to talk of the foul smell emanating from them. Water pollution can even render the water unfit for industrial or agricultural purposes, not alone for drinking. Encroachments formed on the water bodies have led to drastic shrinking of the total area. An example of this in India is the Anchar Lake that has turned into a marsh. River Jhelum has been turned into a drain due to solid wastes and effluents entering into this water body. Its fish population is diseased. Dal Lake of Kashmie can be nicknamed as 'a polluted pond'.

CONCLUSIONS

Pesticides are often considered a quick, easy, and inexpensive solution for controlling weeds and insect pests in urban landscapes. However, pesticide use comes at a significant cost. Pesticides have contaminated almost every part of our environment as pesticide residues are found in soil and air, and in surface and groundwater across the nation, and urban pesticide uses contribute to the problem. Pesticide contamination poses significant risks to the environment and non-target organisms ranging from beneficial soil microorganisms to insects, plants, fish, and birds. Contrary to common misconceptions, even herbicides can cause harm to the environment. In fact, weed killers can be especially problematic because they are used in relatively large volumes. The best way to reduce pesticide contamination (and the harm it causes) in our environment is for all of us to do our part to use safer, non-chemical pest control (including weed control) methods. In order to control water pollution by other elements such as sewage or industrial wastes, the effluents should not be allowed to dump into water reservoirs without proper pretreatment. Further, the constant monitoring and analysis of water by appropriate agencies is essential to avoid any kind of water contamination.

ACKNOWLEDGEMENTS

The authors are thankful to Dr. Abhay Kumar Pandey, Department of Biochemistry, University of Allahabad, Allahabad, India for critical reviewing of the manuscript.

REFERENCES

1. P. H. Daniel, "Investing in Tomorrow's Liquid Gold," 19 April 2006. http://finance.yahoo.com/columnist/article/ trenddesk/pp3748

2. K. R. Munkittrick, M. R. Servos, J. L. Parrott, V. Martin, J. H. Carey, P. A. Flett, G. Potashnik and A. Porath, "Dibromochloropropane (DBCP): A 17-year Reassessment of Testicular Function and Reproductive Performance," Journal of Occupational Environment Medicine, vol. 37, No. 11, November 2005, pp. 1287-1292.

3. P. Cocco, "On the Rumors about the Silent Spring. Review of the Scientific Evidence Linking Occupational and Environmental Pesticide Exposure to Endocrine Disrupting Health Effects," Cadernos Saúde Pública, vol. 18, No. 2, 2002, pp. 379-402.

4. C. Massad, F. Entezami, L. Massade, M. Benahmed, F. Olivennes, R. Barouki and S. Hamamah, "How Can Chemical Compounds Alter Human Fertility?" European Journal Obstetrics Gynecology Reproductive Biology, Vol. 100, No. 2, 2002, pp. 127-137.

5. J. L. Cook, P. Baumann, J. A. Jackman and D. Stevenson, "Pesticides Characteristics that Affect Water Quality". http://insects.tamu.edu/extension/bulletins/water/water_01.html

6. E. Straube, S. Straube and W. Straube, "Hormonal Disruption in Humans," In: D. Pimental, J. L. Cook, P. Baumann, J. A Jackmang

and D. Stevenson Ed., Encyclopedia of Pest Managaement, College Station, 2003.

7.   A. Gupta, D. K. Rai, R. S. Pandey and B. Sharma, "Analysis of some Heavy Metals in Riverine Water, Sediments and Fish from River Ganges at Allahabad," Environmental Monitoring and Assessment, In Press.

8.   Clean Water Act, Section 502, General Definitions (14). http://www.epa.gov/wetlands/regs/sec502.html

9.   S. O. Igbedioh, "Effects of Agricultural Pesticides on Humans, Animals and Higher Plants in Developing Countries," Archives of Environmental Health, vol. 46, 1991, pp. 218-223.

10. J. Jeyaratnam, "Health Problems of Pesticide Usage in the Third World," British Medical Journal, Vol. 42, 1985, pp. 505-506.

11. H. N. Saiyed, V. K. Bhatnagar and R. Kashyap, "Impact of Pesticide Use in India Electronic Journals: Asia Pacific Newsletter: 1999-2003.   http://www.ttl.fi/Internet/   English/Infotion/ Electronic+journals/Asian-Pacific+Newsletter/1999-03/05.htm

12. C. O. Karunakaran, "The Kerala Food Poisoning," Journal of Indian Medical Associtaion, vol. 31, 1958, pp 204- 207.

13. Wadhwani and I. J. Lall, "Indian Council of Agricultural Research," New Delhi, 1972, pp. 44-49.

14. R. Carlson, "Silent Spring," Houghton-Mifflin Co., Boston, 1962.

15. R. A. Liroff, "Balancing Risks of DDT and Malaria in the Global POPs Treaty," Pesticide Safety News, vol. 4, 2000, pp. 3-7.

16. H. L. Bradlow, D. L. Davis, G. Lin, D. Sepkovic and R. Tiwari, "Effects of Pesticides on the Ratio of 16 Alpha/2-Hydroxyestrone: A Biologic Marker of Breast Cancer Risk," Environmental Health Perspectives, vol. 103, 1995, pp. 147-150.

17. M. C. R. Alavanja, J. A. Hoppin and F. Kamel, "Health Effects of Chronic Pesticide Exposure: Cancer and Neurotoxicity," Annual Review of Public Health, vol. 25, 2004, pp. 155-197.

18. M. J. Perry, "Effects of Environmental and Occupational Pesticide Exposure on Human Sperm: A Systematic Review," Human Reproduction Update, vol. 14, 2008, pp. 233-242.

19. "Farm Chemicals Handbook'95," Meister Publishing Co., Willoughby, 1995.

20. R. E. Gosselin, H. C. Hodge, R. P. Smith and M. N. Gleason, "Chemical Toxicity of Chemical Products," The Wilkins & Wilkins Co., Baltimore, 1976.

21. United States Environmental Protection Agency, "Pesticide Industry Sales and Usage, 1990 and 1991 Market Estimates," United States Environmental Protection Agency Publishing, Washington, D.C., 1992.

22. Ontario Ministry of Agriculture and Food, "Grower Pesticide Safety Course," Toronto, 1991.

23. D. Calamari and D. U. Barg, "Hazard Assessment of Agricultural Chemicals by Simple Simulation Models," Prevention of Water Pollution by Agriculture and Related Activities: Proceedings of the FAO Expert Consultation, Santiago, 20-23 October 1992, pp. 207-222.

24. United Nations Environment Programme, "The Aral Sea: Diagnostic Study for the Development of an Action Plan for the Conservation of the Aral Sea," Nairobi, 1993.

25. World Health Organization, "Guidelines for DrinkingWater Quality, Volume 1: Recommendations," 2nd Edition, Geneva, 1993.

26. Y. J. Wang and J. K. Lin, "Estimation of Selected Phenols in Drinking Water with in Situ Acetylation and Study on the DNA Damaging Properties of Polychlorinated Phenols," Archives of Environmental Contamination and Toxicology, vol. 28, 1995, pp. 537-542.

27. S. R. Baker, "The Effects of Pesticides on Human Health," In: C. F. Wilkinson Ed., Advances in Modern Environmental Toxicology, 1990.

28. M. Margni, D. Rossier, P. Crettaz and O. Jolliet, "Life Cycle Impact Assessment of Pesticides on Human Health and Ecosystems," Agriculture, Ecosystems and Environment, vol. 93, No. 1-3, December 2002, pp. 379-392.

29. G. R. Hallberg, "Pesticide Pollution of Groundwater in the Humid United States," Agrigulture, Ecosystem and Environment, vol. 26, No. 3-4, October 1989, pp. 299- 367.

30. McConnell, et al., "Health Hazard Evaluation Report in Pesticides in the Diets of Infants and Children," Pesticides in the Diets of

Infants and Children, National Academy Press, Washington, D.C., 1993.

31.   Oranskey, et al., Seizures Temporally Associated with the Use of DEET Insect Repellent—New York and Connecticut," Pesticides in the Diets of Infants and Children, National Academy Press, Washington, D.C., 1993.

32.   Z. S. Hoar, A. Blair, et al., "Agricultural Herbicide Use and Risk of Lymphoma and Soft Tissue Sarcoma," Journal of the American Medical Association, vol. 256, 1886, pp. 1141-1147.

33.   "Science Daily," 12 March 2009. http://www.science daily.com/releases/2009/03/090306084639.htm

34.   P. B. Singh and V. Singh, "Pesticide Bioaccumulation and Plasma Sex Steroids in Fishes during Breeding Phase from North India," Environmental Toxicology and Pharmacology, Vol. 25, No. 3, May 2008, pp. 342-350.

35.   B. Goldar and N. Banerjee, "Impact of Informal Regulation of Pollution on Water Quality in Rivers in India," Journal of Environmental Management, Vol. 73, No. 2, November 2004, pp. 117-130.

Citations

CHAPTER 1

N. Sethy, R. Tripathi, V. Jha, S. Sahoo, A. Shukla and V. Puranik, "Assessment of Natural Uranium in the Ground Water around Jaduguda Uranium Mining Complex, India," Journal of Environmental Protection, Vol. 2 No. 7, 2011, pp. 1002-1007. doi: 10.4236/jep.2011.27115.

CHAPTER 2

P. Dixit, B. Kar, P. Chattopadhyay and C. Panda, "Seasonal Variation of the Physicochemical Properties of Water Samples in Mahanadi Estuary, East Coast of India," Journal of Environmental Protection, Vol. 4 No. 8, 2013, pp. 843-848. doi: 10.4236/jep.2013.48098.

CHAPTER 3

Liu, Y. , Liu, Q. , Jin, Z. , Cai, L. and Cui, X. (2014) A Simulation Study of Support Break-Off and Water Inrush during Mining under the High Confined and Thick Unconsolidated Aquifer. *Open Journal of Geology*, 4, 599-611. doi: 10.4236/ojg.2014.412044.

CHAPTER 4

E. Ali, S. Muyibi, H. Salleh, M. Alam and M. Salleh, "Production of Natural Coagulant from Moringa Oleifera Seed for Application in Treatment of Low Turbidity Water," Journal of Water Resource and Protection, Vol. 2 No. 3, 2010, pp. 259-266. doi: 10.4236/jwarp.2010.23030.

CHAPTER 5

P. Gbolo and D. López, "Chemical and Geological Control on Surface Water within the Shade River Watershed in Southeastern Ohio," *Journal of Environmental Protection*, Vol. 4 No. 1, 2013, pp. 1-11. doi:10.4236/jep.2013.41001.

CHAPTER 6

Gesicki, A. and Sindico, F. (2014) The Environmental Dimension of Groundwater in Brazil: Conflicts between Mineral Water and Water Resource Management. *Journal of Water Resource and Protection*, 6, 1533-1545. doi: 10.4236/jwarp.2014.616140.

CHAPTER 7

Usha Damodharan (2013). Bioaccumulation of Heavy Metals in Contaminated River Water-Uppanar, Cuddalore, South East Coast of India, Perspectives in Water Pollution, Dr. Imran Ahmad Dar (Ed.), ISBN: 978-953-51-1076-7, InTech, DOI: 10.5772/53374.

CHAPTER 8

Gjessing, E. (2013) Coloured materials in surface water in the sub Arctic Zone: An overview of its formation, properties and environmental changes. *Natural Science*, 5, 400-410. doi: 10.4236/ns.2013.53053.

CHAPTER 9

A. Agrawal, R. Pandey and B. Sharma, "Water Pollution with Special Reference to Pesticide Contamination in India," Journal of Water Resource and Protection, Vol. 2 No. 5, 2010, pp. 432-448. doi: 10.4236/jwarp.2010.25050.

Index